Durability of
Building Sealants

Durability of Building Sealants

Proceedings of the International RILEM Symposium
on Durability of Building Sealants,
Building Research Establishment, Garston, UK

11–12 October, 1994

EDITED BY

J.C Beech

*Formerly Head of Flat Roofs and Sealants Section,
Building Research Establishment,
Garston, UK*

A.T. Wolf

*Global Technology Manager,
Sealant Business Dow Corning Corporation,
Michigan, USA*

Routledge
Taylor & Francis Group

LONDON AND NEW YORK

First published 1996 by E & FN Spon
First edition 1996

Published 2020 by Routledge
2 Park Square, Milton Park, Abingdon, Oxon OX14 4RN
52 Vanderbilt Avenue, New York, NY 10017, USA

First issued in paperback 2020

Routledge is an imprint of the Taylor & Francis Group, an informa business

A catalogue record for this book is available from the British Library

Library of Congress Cataloging-in-Publication data available

ISBN 13: 978-0-367-57966-1 (pbk)
ISBN 13: 978-0-419-21070-2 (hbk)

Contents

RILEM Technical Committee 139-DBS Members

A.T. Wolf (Chairman) Dow Corning Corporation, Michigan, USA.
T. Lee (Secretary) University of Warwick Science Park, Coventry, UK.
J.L. Beasley, Building Research Establishment, Garston, UK.
T. Boettger, Universitaet Leipzig, Leipzig, Germany.
H. Bolte, Universitaet Leipzig, Leipzig, Germany.
E. Brandt, SBI, Horsholm, Denmark.
P.G. Burstroem, Lund Institute of Technology, Lund, Sweden.
L. Crewdson, South Florida Test Service, Miami, USA.
V. Gutowski, CSIRO, Highett, Australia.
S.A. Hurley, Taywood Engineering, Southall, UK.
J. Klosowski, Dow Corning Corporation, Michigan, USA.
M.A. Lacasse, National Research Council of Canada, Ottowa, Canada.
S. Linde, Zel-Aaren Innovation AB, Boras, Sweden.
J.C. Marechal, Centre Scientifique et Technique du Batiment, Grenoble, France.
A. Pagliuca, Oxford Brookes University, Oxford, UK.
T. Wilkins, Forsoc Ltd., Birmingham, UK.

Corresponding members

T. O'Connor, Smith, Hinchman & Grylls Associates, Detroit, USA.
I. Gertis, Lehrstuhl fuer Konstruktive Bauphysik, Stuttgart, Germany.
M. Puterman, Israel Institute of Technology, Haifa, Israel.
W. Sharman, BRANZ, Porirua, New Zealand.

Rilem Secretariat General

M. Brusin, Rilem Secretariat General, ENS, Cachan Cedex, France.

Preface

This book contains the proceedings of the First International Symposium on Durability of Building Sealants which was held October 11-12, 1994, in Garston, England, under the joint auspices of the British Building Research Establishment (BRE) and the International Union of Testing and Research Laboratories for Materials and Structures (RILEM) Technical Committee TC139-DBS (Durability of Building Sealants).

While there is an increasing number of papers published on the durability of sealants, most of these publications are highly focused on either a specific material or a specific weathering protocol. There is, therefore, a need for a publication that more accurately represents the current state-of-the art. In order to fulfil this need the First International Symposium on Durability of Building Sealants was organised.

The symposium was attended by 34 scientists from Europe, USA, and Japan. Nine papers were presented by researchers from USA, Canada, United Kingdom, Germany, Switzerland, and Japan. The main focus of the symposium was on the following topics:

- correlation of changes in sealant properties produced by laboratory ageing techniques and those observed in naturally weathered samples
- prediction of long-term sealant performance from accelerated weathering
- validation of laboratory test methods for prediction of the durability of building joint sealants
- comparison of various short-term ageing methods
- test methods to determine the fatigue resistance of sealants
- review of the state-of-the-art in sealant durability
- contribution of research to the standardisation of sealants

While RILEM does not directly sponsor research, its intent is to act as a catalyst for future research by bringing experts within a field together. The RILEM Technical Committee 139-DBS "Durability of Building Sealants" was inaugurated in July 1991 and has set itself the following objectives:

- to review the present knowledge regarding the assessment of the durability of sealants
- to promote research in this field
- to make recommendations for suitable experimental methods

In particular, the technical committee intends to collect experimental data on the natural and accelerated ageing of sealants and to attempt correlating data obtained according to different ageing regimes using physical performance and chemical analysis methods.

Since the inauguration of RILEM TC139-DBS, its members had the opportunity to meet eight times and discuss recent progress made in their field of expertise. While much progress has been made in recent years in our understanding of the durability of sealants, the discussions within RILEM TC139-DBS most noticeably highlighted the need for a scientific model that would allow the reliable prediction of the in-service performance of sealants. In particular, such a scientific model should be able to link the sealant's initial material properties, such as tensile strength, elasticity or shear modulus and the exposure conditions the sealant experiences during its service life to the sealant's expected service life, given certain "degradation indices" for adhesion and cohesion which are specific for the sealant formulation or the generic sealant category. In order to develop such a scientific model, the durability of various sealants under artificial and outdoor weathering conditions needs to be studied over prolonged periods of time and with a consistent protocol.

The need to generate consistent data on the long-term durability of sealants has led the CSIRO, Australia, and RILEM TC139-DBS to jointly propose an ambitious research project, which plans to expose ten gun-grade sealants (silicone, polysulphide, polyurethane, silicon modified polyether) and one pre-formed tape sealant (acrylic) over a period of ten years at seven exposure sites located in Australia, Michigan (USA), Arizona (USA), Ottawa (Canada), Singapore, England, and New Zealand. The sealants will undergo imposed mechanical cycling by manual adjustment of the width of the specimens twice a year. The sealant samples will be inspected twice annually and analysed using thermo-mechanical and chemical analyses. The outdoor exposure will commence in August 1995.

We hope that the contributions made by nine distinguished researchers, collected in this volume, will stimulate new ideas and initiate further research into the durability of building sealants.

Finally, our special thanks go to the Building Research Establishment, the session chairmen, the authors, and the Chapman & Hall Staff who made the symposium and this resulting publication possible.

Andreas T. Wolf
Current Chairman of
RILEM TC139-DBS

Midland, Michigan, U.S.A.
March 22, 1995

John Beech
Former Chairman (1991-1994) of
RILEM TC139-DBS

Thame, Oxfordshire, U.K.
March 25, 1995

1 STATIC AND DYNAMIC CUT GROWTH FATIGUE CHARACTERISTICS OF SILICONE-BASED ELASTOMERIC SEALANTS

M.A. LACASSE, J.C. MARGESON
and B.A. DICK
National Research Council of Canada, Institute for
Research in Construction, Ottawa, Ontario, Canada

Abstract
Elastomeric based sealants, used to seal moving joints on the exterior of buildings, undergo diurnal cycles of movement superimposed on a yearly seasonal cycle. Hence, they are subject to movements that repeatedly extend and compress the sealant, and this cyclic movement may eventually cause the rupture of the seal due to fatigue of the elastomer. The fatigue resistance of elastomers has been related to that of their tearing resistance and cut growth characteristics. In this study, the dynamic cut growth characteristics of silicone-based elastomers were evaluated to gain an understanding of the phenomena of cyclic fatigue in these types of products and its relation to the durability of sealed joints. Two sets of sealants were cast, each having thin film specimens into which cuts has been intruded at one edge of the film. The first set was used to establish the relationship between the rate of deformation movement and the strain energy density within the specimen. The second set was subjected to cyclic testing from which the intrinsic dynamic cut growth parameters were obtained. Analysis of the cut growth characteristics was undertaken using strain-fatigue life plots from which the durability of the elastomers could be ascertained. Results indicate that the methods can potentially be used to assess the service life of sealant materials and provide a basis for developing an accelerated fatigue test for sealants.
Keywords: cut growth rate, fatigue testing, sealant fatigue behaviour, sealant service life performance, sealants, silicone.

1 Introduction

Expansion joints are typically used to accommodate the cyclic movement of adjacent building cladding elements induced by diurnal and seasonal temperature fluctuations. Elastomeric substances, referred to as sealants, seal these joints and thus prevent the passage of air, moisture and particulate matter through the building envelope while withstanding the cyclic deformation imposed at joints by adjacent elements. Though most sealants in current use are effective at maintaining their integrity throughout numerous cycles, cyclic deformation is considered to be one of the most important factors governing the length of their service-life. The fatigue life of these materials is of primary concern to the construction industry due to the extensive costs involved in replacing failed seals and repairing the damage caused by water infiltration of the building envelope.

Few studies [1-2] have been undertaken to evaluate the fatigue life of sealants products and in which the fatigue life of different sealants under diverse climatic conditions has been characterized.   However, because of their elastomeric properties, it is expected that the fatigue life of sealants can be described using, as an initial basis, derivations of the work undertaken on the fatigue of natural rubber compounds.   Predominately, the studies into rubber fatigue have dealt with the concept of crack growth which is widely felt to be the primary mechanism for cohesive failure of the material under cyclic stress.

Rivlin and Thomas [3] first described cut growth theory for elastomers that gave rise to the concept of a characteristic tearing energy, demonstrated to be an intrinsic property of elastomeric materials, in which the tear criterion is given by:

$$-\left(\frac{\partial W}{\partial c}\right)_l = T \bullet t \tag{1}$$

where c is the length of the cut, t is the thickness of the test piece, T is the tear energy parameter, and W is the stored strain energy in the specimen. In (1), l indicates that the differentiation is to be carried out under constant displacement of the boundaries of applied force.

From (1), the cut growth at a constant strain and cut length is observed to be independent of the size of the initial cut. Additionally, since the characteristic tearing energy is solely a consequence of energy flow, when the cut growth behaviour is expressed in terms of T, the dimensions of the test piece, and the method by which the tearing energy parameter was obtained are irrelevant as descriptive parameters. Consequently, the characteristic tearing energy can be seen as a purely generic description for the cut growth in a particular material.

Greensmith [4] showed that the relationship between strain energy and tear energy was extendible beyond the initial cut propagation to the extended portion of the tear energy cut growth curve. Lake and Lindley [5] expanded upon this theory and showed the relationship between these properties could be described according to the following relationship:

$$T = 2K \cdot (\lambda) \cdot W_o \cdot c \tag{2}$$

where W_o is the elastic energy density in the specimen, and $K(\lambda)$ is a proportionality constant depend on the strain [4].

Lake and Lindley [5] showed that the tear energy T was dependent on the change in cut length differentiated with respect to the number of cycles the specimen was subjected to. This relationship is shown in Figure 1.

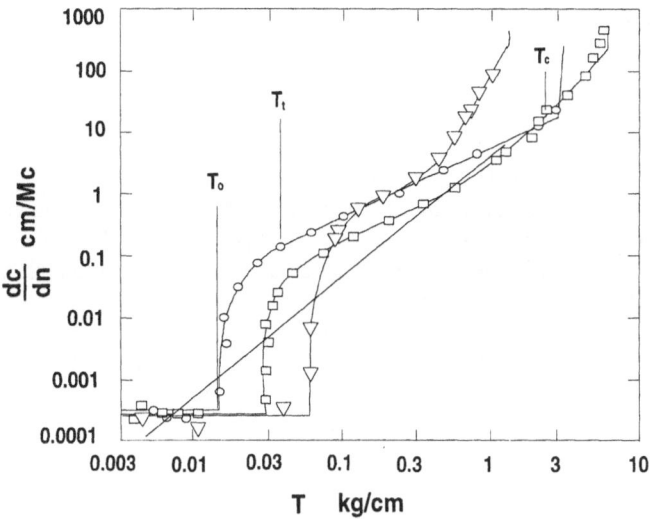

Fig. 1. Cut growth rate *dc/dn* vs. tearing energy *T* for three natural gum vulcanizates showing the minimum, T_0, transitional, T_t, and catastrophic, T_c, tearing energy events according to Lake and Lindley [5].

Lake identified three critical events on the tear energy curve. He defined these as T_0 the minimum energy necessary to initiate cut growth, T_t the point at which the cut growth changed from a linear to a square growth rate, and T_c where the cut growth became catastrophic. The four regions have been similarly described by Fukahori [6] who denoted each of the four regions by the parameter β.

$T \leq T_o$	β_1
$T_o \leq T \geq T_t$	β_2
$T_t \leq T \geq T_c$	β_3
$T \geq T_c$	β_4

The β_3 region on the cut growth rate curve was the area of focus for much of the early work which was done on this phenomenon [7-10]. Gent et al. [7] were able to correlate the relationship of cut growth and tear energy to that of fatigue life within the β_3 region according to the following relation:

$$\frac{dc}{dn} = \frac{1}{G} T^\beta$$

(3)

in which, G and β are the intrinsic dynamic cut growth constants of the material being evaluated.

If (3) is expressed in the following format:

$$\log \frac{dc}{dn} = \beta \, \log T + \log G \tag{4}$$

then the values for β and G can be derived a plot of the $\log dc/dn$ - $\log T$ curve.

Initially, examination of (3) indicates that it may be used to derive an expression which characterizes the fatigue life of a sealant specimen [6,10]. In fact, if (3) is solved for dn and then integrated between an initial and final cut length, it can be shown [6,10] that,

$$n = \int_{c_o}^{c} \frac{G}{\left(2K(\lambda) \cdot W_o\right)^{\beta}} dc = \frac{G}{(\beta - 1) \cdot \left(2K(\lambda) \cdot W_o\right)^{\beta}} \left[\frac{1}{c_o^{\beta - 1}} - \frac{1}{c^{\beta - 1}} \right] \tag{5}$$

In (5), the expression for tearing energy (see (2)) is assumed. Further, if $c \gg c_o$, as would be expected at failure, (5) can be simplified to [6,10]:

$$N = \frac{G}{(\beta - 1) \cdot \left(2K(\lambda) \cdot W_o\right)^{\beta}} \cdot \frac{1}{c_o^{\beta - 1}} \tag{6}$$

where N is the number of cycles to failure.

In this study, it is proposed to evaluate the cut growth parameters, β and G, of a commercially available silicone sealant and based on these parameters, determine the fatigue life of these products. That is, for experimentally derived values of the cut growth parameter, the number of cycles to failure, N can then be estimated from (6) based on given values of cut length, c_o, and extension ratio, λ, respectively.

Two sets of specimens were prepared, each set having predetermined cut lengths intruded along the length of their long axis. One set was subjected to tensile testing in order to obtain a representation for the strain energy density at various levels of strains, whereas the second set was cycled to predetermined maximum strain levels with the cut growth rate measured with respect to cycle count.

2 Experimental

2.1 Materials
A silicone-based one-part moisture cured, gun-grade, commercial product of medium gray colour was used in this study. Sealant specimens were cast in thin sheets of 70 x 127 x 1.50 mm size and cured in laboratory conditions (21 ± 2°C and 50 ± 4% RH) for 14 days. After curing, the sheets were cut into strips of 25.4 x 127 mm.

Two series of sealant specimens were prepared for this study: the first set, comprised of 30 specimens, was subjected to tensile strain in order to obtain a value for the strain energy density at various strain ratios; the second series (104 specimens) was cycled to predetermined maximum strains with the cut growth rate measured as a function of the number of cycles. In each test series, specimens had cuts of nominally predetermined lengths intruded using a razor blade. These cuts were made perpendicular to the longitudinal axis at mid-way along the length of the specimens, as illustrated in Figure 2.

Actual cut length measurements were made from a magnified image of the specimen shown on a television monitor and obtained from a CCD video camera. Using this technique, the effective magnification of the cut length was approximately 30 times actual.

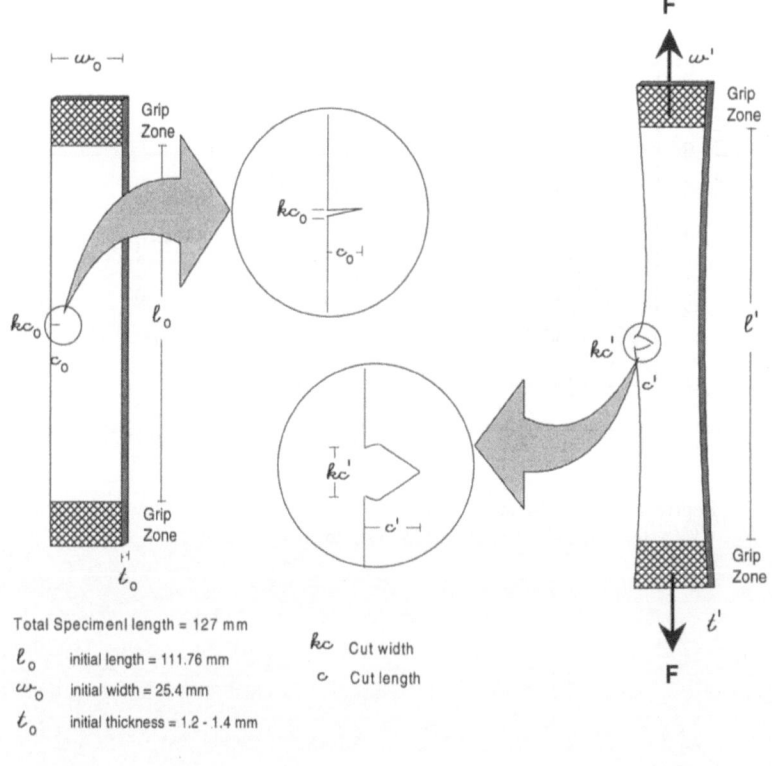

Total Specimen length = 127 mm

ℓ_o initial length = 111.76 mm

w_o initial width = 25.4 mm

t_o initial thickness = 1.2 - 1.4 mm

kc Cut width

c Cut length

Initial Specimen Dimensions   **Specimen under load**

Fig. 2. Specimen used in the static and dynamic cut growth studies showing both undeformed and deformed specimens.

2.2 Procedure - *Strain energy density*

A set of three (3) sealant specimens was subjected to tensile extension according to the schedule shown in Table 1. Tests were conducted at 21°C and 50 % RH using an Instron Universal testing machine. Specimens having different lengths of intruded cuts were extended to 90% of the initial gauge length (l_0 = 111.76 mm). For each test, values for the strain energy density at a given extension were obtained from an integration of the load-displacement curve up to the appropriate % extension. A representative load-displacement curve for a sealant specimen having a 3 mm intruded cut length is shown in Figure 3. Results are provided for the strain energy determined at 45 and 90% extension of the specimen.

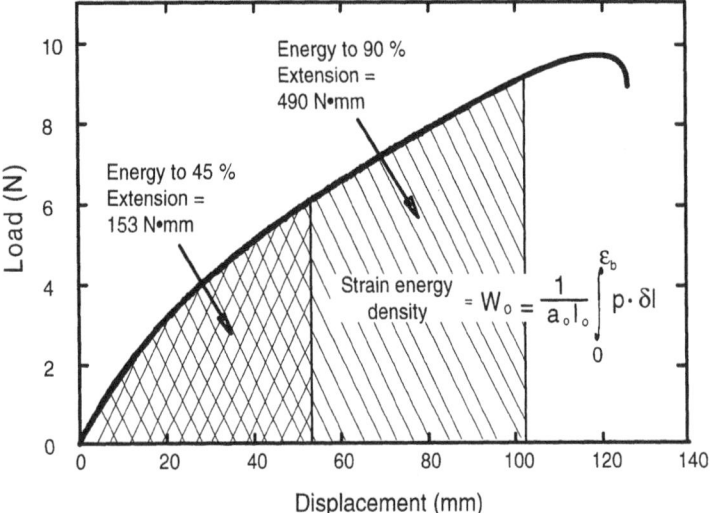

Fig. 3. Representative load-displacement curve showing values of strain energy at 45 and 90% extension and the relationship to strain energy density.

Table 1. Evaluation of Strain Energy Density: Number of test specimens according to specimen characteristics and test speeds.

Cut Length (mm)	Crosshead speed (mm/min)	
	10	1000
1.0	3	3
2.0	3	3
3.0	3	3
4.0	3	3
5.0	3	3

2.3 Procedure - *Dynamic cut growth*

A jig was used to simultaneously subject a series of four (4) specimens to cyclic testing. The jig was fabricated such that a set of two transverse bars could be mounted into the top and bottom test grips of a servo-hydraulic material testing system (MTS). Onto these bars were placed four pairs of individual specimen grips, each placed at intervals of 12.7 mm along the length of the bar. Specimens placed in the grips were tightened sufficiently to prevent slippage with care being taken not to damage the sealant specimen in the process. Slippage was monitored by marking the position of the grip edge with a felt tip marker before the testing routine commenced.

Hence, as shown in Table 2, a series of 26 cyclic tests was carried out, each series consisting of a set of four sealant specimens having different intruded cut lengths and being subjected to tensile extension in amplitudes ranging from 30 to 90% of gauge length.

The cyclic test program was undertaken using a MTS programmed to cycle specimens at 0.5 Hz. The time-displacement curve was split into two equal portions: the first segment consisted of the positive half of a sinusoidal curve while the latter portion held the sample in a relaxed state. The characteristic time-displacement curve is shown in Figure 4.

Table 2. Test Schedule for specimens subjected to cyclic testing at different extension amplitudes (%).

Cut Length (mm)	30%	40%	45%	50%	55%	60%	70%	80%	90%
1.0	0	0	0	0	0	4	4	4	4
2.0	0	0	4	4	4	4	4	4	4
3.0	0	4	4	4	4	4	4	4	0
4.0	4	4	4	4	4	4	4	4	0

Each cycle began at Point A (see Figure 4) on the sinusoidal curve, the number cycles being recorded by the MTS. Measurements of the cut progression were made by halting the cycling procedure at an integer cycle and using a CCD video camera to capture an image of the cut.

In order to facilitate the accurate measurement of cut length with the CCD video camera, a small load was induced onto the specimen when the cut length measurement was being made. Cut progression in the specimens were recorded, using a magnified image from a CCD camera, to a VHS format video cassette. The magnification factor was determined by placing circular metallic discs of known size into the camera's view and comparing the size of the observed image with the actual size of the disc. Cut lengths were measured directly from the monitor using digital calipers having a resolution of ±0.01 mm. Using this technique, the original image was magnified 30 X thus allowing the cut length to be measured with an accuracy of ± 0.003 mm.

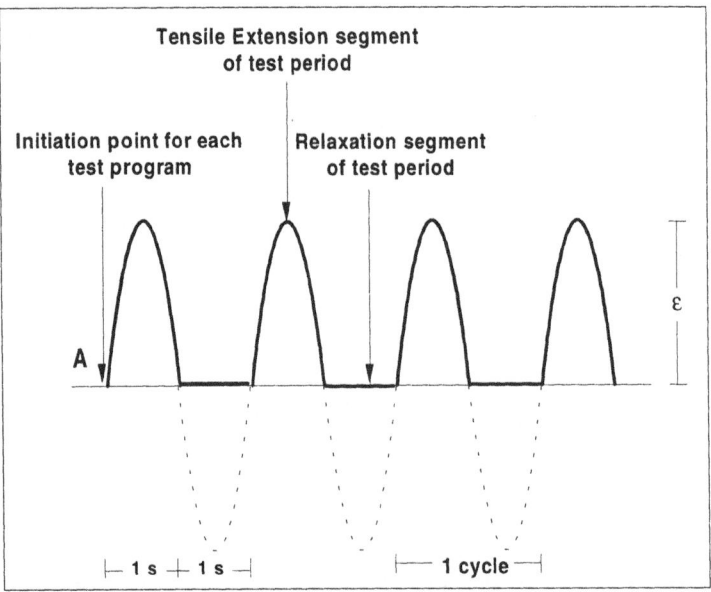

Fig. 4. Dynamic Cut Growth Cyclic Displacement Input from Nominal Gage Separation.

3 Results and Discussion

3.1 Static Cut Growth

Average values for strain energy density, W_o, calculated according to the relationship given in Figure 3, are provided in Table 4.

Table 4 Values of W_o (N/mm^2) as a function of intruded cut length and extension ratio.

Ext %	Intruded Cut length (mm)				
	1	2	3	4	5
10	0.00018	0.00015	0.00024	0.00016	0.00021
20	0.00066	0.00053	0.00087	0.00060	0.00077
30	0.00136	0.00110	0.00179	0.00125	0.00161
40	0.00223	0.00182	0.00295	0.00210	0.00267
45	0.00274	0.00224	0.00361	0.00258	0.00326
50	0.00328	0.00269	0.00430	0.00311	0.00391
55	0.00386	0.00318	0.00506	0.00366	0.00461
60	0.00448	0.00372	0.00586	0.00427	0.00534
70	0.00583	0.00489	0.00759	0.00556	0.00693
80	0.00730	0.00620	0.00947	0.00699	0.00865
90	0.00755	0.00766	0.01150	0.00853	0.01049

Values of the characteristic tearing energy, T, were calculated for different cut lengths and extensions using the relationship provided by Lake and Lindley [3]: $T = 2K \cdot (\lambda) \cdot W_0 \cdot c$, and values of $K(\lambda)$ derived from Greensmith [2], the later provided below in Table 5.

Table 5. Values for $K(\lambda)$ at each extension ratio from Greensmith [2]

λ	$K(\lambda)$
1.3	2.65
1.4	2.45
1.45	2.33
1.5	2.28
1.55	2.26
1.6	2.35
1.7	2.2
1.8	2.15
1.9	2.05

3.2 Dynamic Cut Growth

A representative result from cut growth test of specimens subjected to cyclic testing is shown in Figure 5, in which the cut growth (mm) is plotted as a function of the number of cycles for a set of four specimens (Sp 1 to Sp 4) having a 3mm intruded cut length and subjected to 45% extension. Values of the change in cut growth in relation to the number of cycles (i.e. dc / dn) were derived by simply calculating the difference in cut length in relation to the corresponding interval between observations.

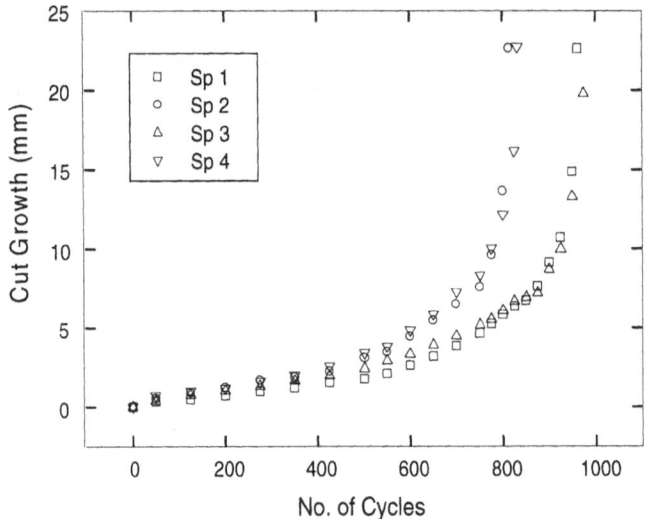

Fig. 5. Cut growth (mm) as a function of the number of cycles tested for four specimens having an initial cut length of 3 mm and subjected to a 45 % extension.

The cut growth rate can plotted as a function of the tearing energy, T, and a representative curve is shown in Figure 6 for a specimen having an intruded cut length of 3 mm and subjected to a 45% extension. A result comparable to that obtained by Lake and Lindley (Figure 1 [5]) is obtained when the relationship provided in Figure 6 is plotted in a log-log format, as shown in Figure 7.

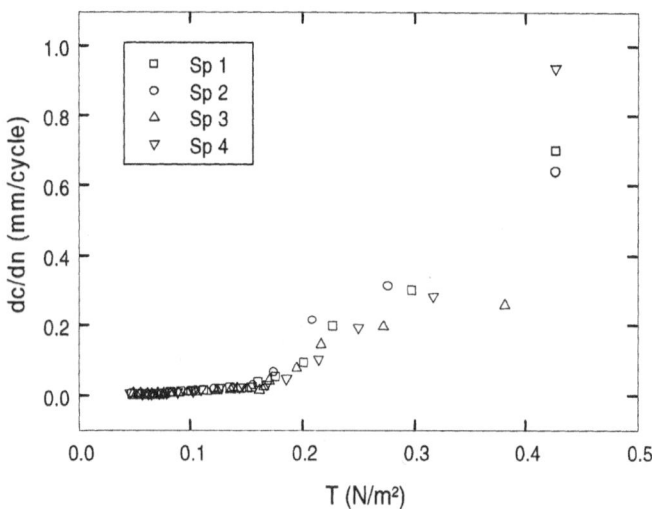

Fig. 6. Cut Growth Rate (dc/dn) as a function of tear energy (T) for the four specimens with an initial cut length of 3 mm and a 45 % extension.

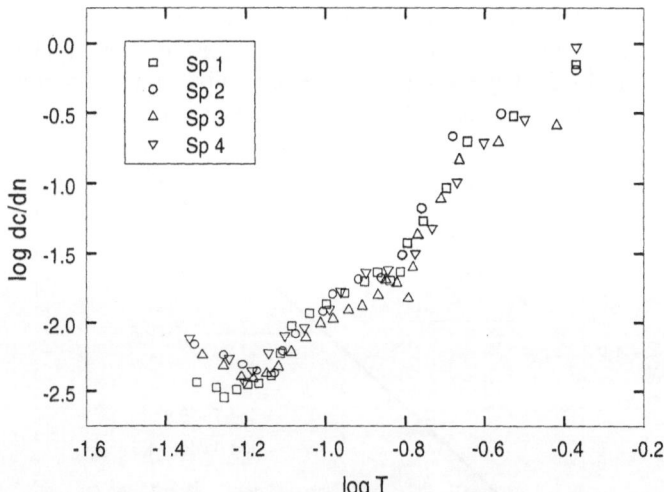

Fig. 7. Log cut growth rate (dc/dn) as a function of log tear energy (T) for specimens having an intruded cut length of 3 mm and subjected to 45 % extension.

A generalized curve based on results of cut growth tests on a specimen having an intruded cut of 3 mm and subjected to 45% extension is shown in Figure 8. Also indicated on the figure are the various regions (β_1, β_2, β_3, and β_4) of cut growth as characterized earlier in Figure 1. Qualitatively, it appears that sealant materials can be characterized using this technique as has been done for other elastomer materials. Based on these results, values for the intrinsic dynamic cut growth constants, β and G, can be calculated from the β_3 portion of the curve.

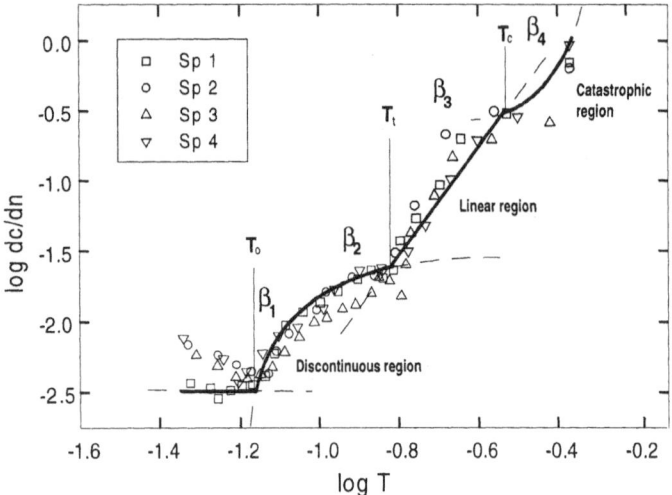

Fig. 8. Generalized curve for sealant specimens with an initial cut length of 3 mm and a 45 % extension.

The linear region represents the area of continued cut growth, hence as shown in Figure 9, values for β and G can readily be derived from a plot of (4) ($\log dc/dn = \beta \log T - \log G$). The plot is representative of a specimen having an intruded cut length of 3 mm and subjected to 45% extension.

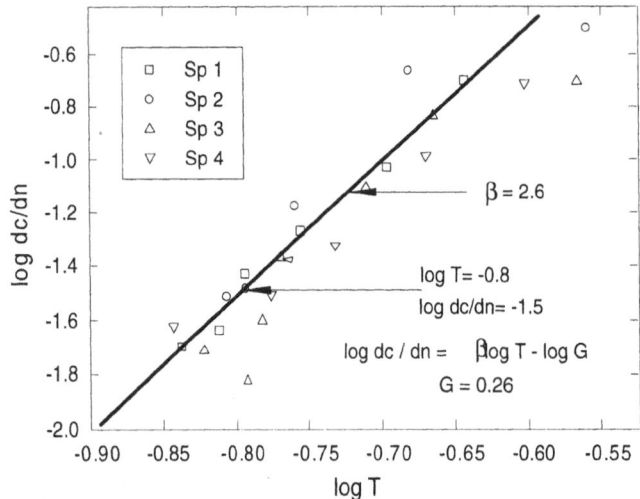

Fig. 9. β_3 portion of the log dc/dn - log T curve, showing the slope of the best fit linear regression.

A linear regression of data points provided values of β = 2.6 and G = 0.26.

It is possible to calculate values of cycles to failure, N, based on values of β and G obtained above, for a given intruded cut length, c, using the relationship provided in (6). Thus, given:

$$N = \frac{G}{(\beta - 1) \cdot (2K(\lambda) \cdot W_o)^{\beta}} \cdot \frac{1}{c_o^{\beta-1}}$$

one obtains:

$$N = \frac{[0.26]}{([2.6] - 1) \cdot (2 \cdot [2.33] \cdot [0.00361])^{[2.6]}} \cdot \frac{1}{[3.0]^{[2.6]-1}} = 1150$$

Data from the dynamic cut growth study is plotted in Figure 10, in which the average number of cycles to failure for a given intruded cut length is provided as a function of the % extension.

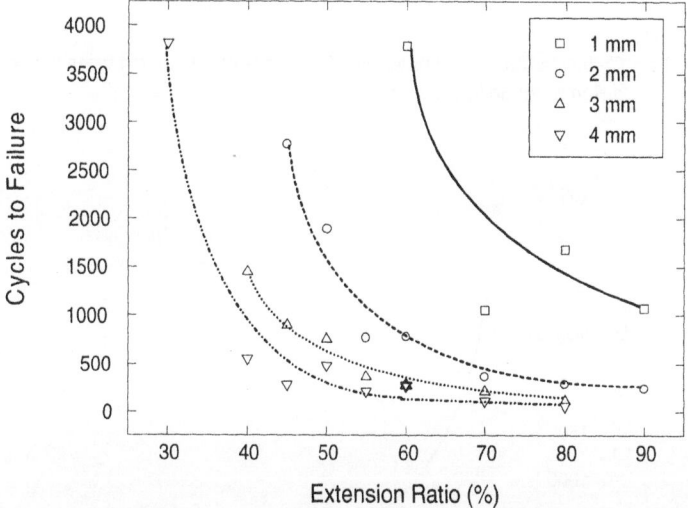

Fig. 10.   Cycles to failure as a function of % extension, derived from experimental data for specimens having different intruded cut lengths.

Similarly, the number of cycles to failure for values of N calculated according to equation (6) for different intruded cut lengths as a function of the % extension is shown in Figure 11. A comparison of the experimental in relation to the calculated results for a specimen having a 4 mm intruded cut length is shown in Figure 12. Collectively these preliminary results appear to indicate that this technique can be used as a basis for evaluating the intrinsic cut growth characteristics of sealant materials, and could also be used to predict the fatigue life of these products.

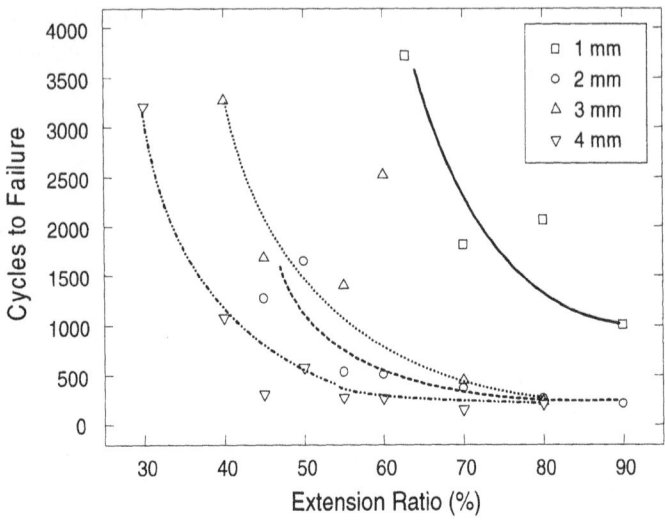

Fig. 11. Cycles to failure as a function of % extension calculated using equation (6) at different intruded cut lengths.

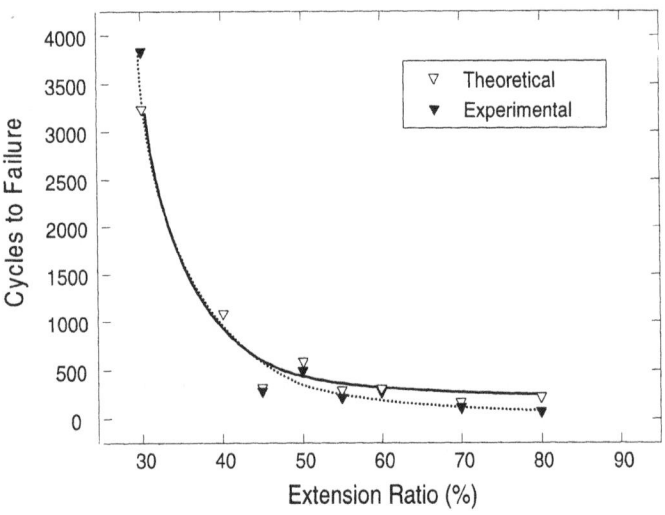

Fig. 12. Comparison of cycles to failure obtained from cyclic tests in relation to those calculated from equation (6) for a specimen having a 4 mm intruded cut length.

4 Summary

This paper summarizes the preliminary work conducted to determine the validity of applying cut growth theory to elastomeric sealants. Analysis has indicated that, in general, cut growth theories, as developed for rubber, can be applied to silicone based elastomeric sealants. Though the tearing energy - cut growth rate relationship for this sealant was shown to be valid, more work is needed directly establish the fatigue life of samples using this technique. Further analysis into the relationship between the cut growth rate and observed fatigue life is required in order to relate the results of fatigue tests using this method and that determined from subjecting model sealant beads to cyclic fatigue. There is also the possibility of testing samples at relatively high strain levels to obtain an accurate representation of the behaviour at smaller amplitudes and initial cut lengths, as is more likely to occur in practice. The test may be accelerated by increasing the initial cut length and/or the strain.

5 References

1. M. A. Lacasse, "Advances in Test Methods to Assess the Long-Term Performance of Sealants", Science and Technology of Building Seals, Sealants, Glazing and Waterproofing: 3rd Volume, ASTM STP 1254, James C. Myers, Ed., American Society for Testing and Materials, Philadelphia, to be published (1994).
2. M. A. Lacasse, J. E. Bryce, J.C. Margeson, "Evaluation of Cyclic Fatigue as a Means of Assessing the Performance of Construction Joint Sealants, Part I: Silicone Sealants", Science and Technology of Building Seals, Sealants, Glazing and Waterproofing: 4th Volume, ASTM STP 1234, David H. Nicastro, Ed., American Society for Testing and Materials, Philadelphia, to be published (1995).
3. Rivlin, R.S., Thomas, A.G., 1953. "Rupture of Rubber. I. Characteristic Energy for Tearing", *Journal of Polymer Science*, 10, p. 291-318.
4. Greensmith, H.W., 1963. "Rupture of Rubber. X. The Change in Stored Energy on Making a Small Cut in a Test Piece Held in Simple Extension", *Journal of Polymer Science*, 7, p. 993-1002.
5. Lake, G.J., Lindley, P.B., 1965. "The Mechanical Fatigue Limit for Rubber", *Journal of Polymer Science*, 9, p. 1233-1251.
6. Fukahori, Y., 1986. "Estimation of Fatigue Life in Elastomers. Theoretical Analysis and Application of S-N Curves with Fracture Mechanics", *International Polymer Science and Technology*, 13-3, p. T/(37-45).
7. Thomas, A.G., 1958. "Rupture of Rubber. V. Cut Growth in Natural Rubber Vulcanizates", *Journal of Polymer Science*, 16, p. 467-480.
8. Gent, A.N., Lindley, P.B., Thomas, A.G., 1964. "Cut Growth and Fatigue of Rubbers. I. The Relationship Between Cut Growth and Fatigue", *Journal of Polymer Science*, 8, p. 455-466.

9. Lake, G.J., Lindley, P.B., 1964. "Cut Growth and Fatigue of Rubbers. II. Experiments on a Noncrystallizing Rubber", *Journal of Polymer Science*, 8, p. 707-721.

10. Royo, J., 1992. "Fatigue Testing of Rubber Materials and Articles", *Polymer Testing* 11, p. 325-344.

2 THE CORRELATION OF MODULUS CHANGES IN BUILDING SEALANTS AFTER ARTIFICIAL AGEING WITH THOSE THAT OCCUR AFTER NATURAL WEATHERING

J.L. BEASLEY
Building Research Establishment, Garston, Watford, UK

ABSTRACT

Nine building sealants have been exposed to natural and artificial weathering. Significant correlations in modulus changes between artificially and naturally weathered, freshly made sealants, have been found. Significant correlations have been found for artificially weathered, well cured, sealant and material that has been exposed to UK and Dubai weather for periods up to 6½ years.

Measurements of modulus values on sealants that have been stored under standard conditions for up to 2 years show that some sealants require over twelve months before stability is achieved.

Keywords:Sealants, durability, weathering, ageing, modulus, curing.

INTRODUCTION

An objective of researching into the durability of sealants is to discover a laboratory based method of accelerating the effects of the natural environment. This can be attempted by exposure to one or more agents that are known to alter the physical properties of sealant materials or their ability to adhere to substrates. Some common methods that can be used are heating, ultra violet radiation, and moisture.

If the sealant is used in a situation where the principal agent of degradation is immediately obvious then the choice of agent to be used in a laboratory study is usually straightforward. Unfortunately reality is unlikely to be so simple and most sealants will be subjected to a number of degrading agents during their service life. The agents used in this programme of work were chosen partly because it is known that they can affect the physical properties of sealants, and partly by the constraint imposed by the apparatus available.

However once the choice of degrading agent is made the properties that will be used to monitor changes must be chosen. Properties that can be easily measured such as modulus or extensibility have the unfortunate attribute of also changing as the state of cure of the sample changes. The test programme was designed so that the effects of changes in the state of cure could be minimized.

This paper is based on laboratory studies of sealant durability using both newly cured [1] and well cured [2] sealants. In addition the effect of the laboratory ageing regimes is compared with that produced by natural weathering exposures of up to 6½ years in the UK, Dubai and Sierra Leone [3].

EXPERIMENTAL PROGRAMME

MATERIALS

Nine sealants that are representative of those available in the UK were used. They were:

Code	Colour	Description
P13	Black	One Part Polysulphide to BS5215 [4]
2P10	Grey	Two Part Polysulphide to BS4254 [5]
2P15	Black	Two Part Polysulphide to BS4254 [5]
S6	Grey	Silicone to BS5889 [6] Type A
S12	Bronze	Silicone to BS5889 [6] Type B
S13	Black	Silicone to BS5889 [6] Type A
U1	Grey	One Part Polyurethane
U3	Grey	Multi Part Epoxy Polyurethane
U6	White	One Part Polyurethane

The test pieces consisted of a 12mm x 12mm x 50mm bead of sealant between one mortar and one mill finished aluminium substrate. All the substrates were primed in accordance with the relevant manufacturer's instructions.

WEATHERING EXPOSURES

Five exposure conditions were used:

Control conditions. Storage at 23±2°C 50% rh

Natural weather. Placed on the exposure site at Garston.

Heat. Stored in a ventilated oven at 70±2°C.

UV and heat. Placed in a QUV artificial weathering machine and subjected to continuous UV(B) radiation and heat at 70°C.

UV and condensation. Placed in a QUV artificial weathering machine and subjected to an 8 hour weathering cycle that consisted of 4 hours of UV(B) radiation at 70°C followed by 4 hours of condensation at 40°C.

The ultra violet radiation was produced by the Q-Panel Company's UVB-313 fluorescent tubes which give a spectrum centred around 313 nm.

METHOD

The experimental programme was designed to enable the effects of continuing cure to be distinguished from those produced by the different ageing procedures used. Each of the sealants used was made up into two batches of test samples. Samples from the first batch were cured for four weeks at 23±2°C and 50±5% rh before being allocated to one of the exposure regimes. The test pieces from the second batch were stored for one year under the same conditions prior to use in the second year.

During the first and second year of the test programme, five replicates of each sealant were removed from each exposure condition at periods of 2, 4, 8, 16, 28, 40 and 52 weeks, prior to tensile testing.

Test pieces were pulled to destruction on a Hounsfield H2000 tensile testing machine. The force at 15mm width was recorded.

Changes in force at 15mm width between naturally weathered and artificially aged sealant were correlated by carrying out a linear regression analysis.

RESULTS AND DISCUSSION

FIRST YEAR [1]

Rate of Cure

Measurement of the force at 15mm width of the control samples in year one, (Table 1) can be used as an indicator of the state of cure of the different sealants. Of interest is the rate at which the force changes during the early and later parts of the year and also the extent to which the sealant achieved the 52 week force value after the initial four week curing period.

In general, the multi component sealants achieve a higher state of cure after the initial four week curing period than the one component systems. The final rate of change of modulus is around an order of magnitude lower than the initial rate of change.

The rate of cure is expressed as a percentage of the initial modulus value per day. The initial rate of change is calculated over a 14 day period. The final rate is calculated over a 168 day period. The calculation assumes a linear rate of cure which is probably not true, but the difference in the two values is a reasonable indicator of the change of rate of cure between the beginning and end of the year.

Correlation of Natural and Laboratory Weathering

The percentage difference in force at 15mm width after 52 weeks in the different exposure conditions from the force after 52 weeks in control conditions is shown in table 2.

Table 1

Changes in force at 15 mm width in control specimens

Mean Force/ newtons at 15mm width	P13	2P10	2P15	S6	S12	S13	U1	U3	U6
Initial	8	60	58	27	25	143	54	37	37
At 2 weeks	46	63	64	35	59	141	67	48	43
At 28 weeks	112	65	74	53	103	149	110	61	67
At 52 weeks	124	73	82	79	105	162	130	67	72
Initial modulus as % of M_{52}	6	83	70	35	24	89	42	54	51
Rate of change (a) %	32	0.3	0.8	2.1	9.6	-.1	1.7	2.2	1.3
Rate of change (b) %	.88	.08	.08	.58	.06	.05	.22	.10	.08

(a) initial rate, (b) final rate, see text.

Table 2

Percentage difference from controls at 52 weeks

Sealant	Natural Weather	Heat 70°C	UV(B) 70°C	UV(B) + Condensation
P13	-29	-17	-41	-50
2P10	-19	+31	+23	+6
2P15	-7	+39	+11	+17
S6	-25	+10	+5	-17
S12	+2	+27	+12	+11
S13	-10	+36	+36	-16
U1	+4	+113	+46	-20
U3	-28	+23	+23	-59
U6	+4	-29	+41	+2

There is no significant relationship, at the 5% level, between the changes due to natural weather after 52 weeks and those produced by similar periods of artificial ageing. However as the climate at the BRE Garston exposure site is relatively mild it is not surprising that it does not produce the same changes over similar periods of time as those produced in more severe laboratory based exposures.

When the changes in modulus produced from shorter periods of exposure to artificial ageing are compared with those from 52 weeks of natural weathering some significant correlations are obtained, Tables 3 and 4.

Table 3

Correlation Coefficients for linear regression:
force at 15mm width. 12 months natural weather with laboratory tests.

Exposure	2 weeks	4 weeks	8 weeks	16 weeks
Heat 70°C	0.726 *	0.706 *	0.713 *	0.607 NS
Heat/UV 70°C	0.735 *	0.725 *	0.708 *	0.724 *
UV + Condensation	0.845 **	0.170 NS	0.915 ***	0.831 **

Key to significance levels :

*** at 0.1% , ** at 1% , * at 5% , NS not significant.

Table 4

Linear regression coefficients for force differences
from controls. 12 months natural weather with
laboratory tests.

Exposures	2 weeks	4 weeks	8 weeks	16 weeks
Heat 70°C	0.879 ***	0.838 **	0.639 *	0.502 NS
Heat/UV 70°C	0.875 ***	0.807 **	0.804 **	0.696 *
UV and Condensation	0.775 **	0.814 **	0.595 NS	0.773 **

Key to significance levels :

*** at 0.1% , ** at 1% , * at 5% , NS not significant.

The correlations shown in table 3 include the effects of both continuing cure and those of weathering whereas, those shown in table 4 can, by looking at differences from same age controls, be attributed more closely to the weathering factors alone. The levels of significance in table 4 are, in general, better than those found in table 3. The two sets of correlations from the first year, Tables 2 and 3, show that the effects of cure and ageing can be differentiated and that in general the correlations between laboratory exposures and natural weathering tend to be of higher significance if the effects of continuing cure are "factored out".

There is little difference in either table 3 or 4 in the levels of significance for the use of heat alone or in conjunction with UV radiation. Given the short periods of time involved and that the effect of the UV radiation is mostly confined to the surface of the samples, this suggests that raising the temperature is the more effective, in the short term at least, of the two degrading agents.

Table 4 shows that the significance of the correlation of heat ageing, and of UV with heat, with 52 weeks of natural weather is highest after 2 weeks of exposure and thereafter the level of significance of the correlation declines. The UV condensation cycle provides correlations that are, in general, consistently good.

The results shown above suggest that it is possible to perform relatively short term ageing procedures on freshly cured sealant that will produce changes in modulus that are correlated with those that will occur under natural conditions after about one year.

SECOND YEAR[2]

Rate of Cure

Table 5 shows how the force values of the control samples varied in the second year of the test programme. The rate of cure in year 1 is the difference in force at 52 weeks from the initial value expressed as a proportion of the 104 week value. The rate of cure in year 2 is the difference in force value between weeks 52 and 104 expressed as a proportion of the 104 week value. It can be seen that a number of the sealants appear to be curing even after one year in control conditions. This is most noticeable in the one part polysulphide sealant P13.

Table 5

Changes in in force at 15mm width
in control specimens during years 1 and 2

Time	P13	2P10	2P15	S6	S12	S13	U1	U3	U6
Initial	8	60	58	27	25	143	54	37	37
52 weeks	124	73	82	79	105	162	130	67	72
104 weeks	144	70	83	85	94	173	142	68	70
Rate of cure in year 1	.81	.18	.29	.61	.85	.11	.53	.45	.50
Rate of cure in year 2	.14	-.05	.01	.06	-.11	.06	.08	.01	-.02
% final modulus: initial	6	83	70	35	24	89	42	54	51
% final modulus: 52 weeks	86	100	99	94	100	94	92	99	100

Correlation of Natural and Laboratory Weathering

When the same correlations of laboratory ageing techniques with 52 weeks of natural weather were carried out on the second year samples, as had been performed on the samples from year one, only one significant relationship was found. This was for the eight weeks of the UV radiation and condensation cycle and was found to be significant at the 1% level. This correlation was significant at the 0.1% level in the first year.

The change in the modulus that occurs when a sealant cures is relatively large when compared to the initial value and happens over a relatively short period of time. The change in modulus that will occur in well cured sealants under natural weathering is probably relatively small in relation to the initial value and can occur over a period of up to twenty years. This suggests that correlations of modulus changes in well cured sealants after laboratory ageing should be carried out with sealants that have been exposed to natural weather for longer than 52 weeks.

COMPARISONS WITH OTHER WEATHERING ENVIRONMENTS[3]

BRE has exposure sites in the UK and overseas at Dubai and Sierra Leone. Sealant test specimens have been exposed to natural weather for up to 15 years on all of these sites. In addition control specimens have been stored under standard conditions, 23±2°C and 50±5% rh, for the same length of time.

Samples have been retrieved at regular intervals and modulus values obtained for the exposed and control specimens. The UK exposure site is subjected to a mild temperate climate in the south of England. The climate in Sierra Leone is hot and humid. The climate in Dubai is hot and dry.

Unfortunately the materials that had been exposed for periods longer than 6½ years were too dissimilar, because of changes in formulation, from those used in the laboratory ageing regimes to make any attempt at correlating changes in modulus values valid. Samples exposed in the UK and Dubai for up to 6½ years and in Sierra Leone for up to four years were essentially the same as those used in the laboratory ageing.

Because the two sets of sealants do not share the same control samples the correlation of changes in modulus was carried out using the difference from the respective control values expressed as a proportion of the control value.

Correlations of changes in modulus, after laboratory ageing, were calculated for both the freshly cured and well cured sealants, with the sealants exposed to periods of natural weather in the UK and overseas. The results are shown in tables 6 and 7.

The results show that the well cured samples tend to show better correlations with longer periods of weather exposure ,Table 6, whereas the freshly cured samples, Table 7, show better correlations with shorter periods of exposure.

The results for the well cured samples, Table 6, show that the use of heat or heat in combination with UV radiation produces negative correlations at the 5% level of significance. The use of the UV and condensation cycle produces significant and very significant correlations for comparisons with Dubai and UK exposure. The most significant level of correlation, 0.1%, occurs between 52 weeks of UV and condensation with 6½ years exposure in the UK. The "acceleration" of change in modulus is in good agreement with that found for freshly cured sealant after 8 weeks of UV and condensation when compared with 52 weeks natural weather, table 3.

Table 6

Summary of significant correlations of % modulus change between natural weathering and well cured laboratory test specimens.

Laboratory Test		Correlation with weathering	
Condition	Duration	Location	Significance Level
UV/Heat	8 weeks	Dubai - 2 years	5%*
UV/Cond	16 weeks	UK - 6½ years	5%
UV/Cond	28 weeks	UK - 6½ year	5%
UV/Cond	40 weeks	UK - 6½ years	1%
UV/Cond	40 weeks	Dubai - 2 years	5%
UV/Cond	40 weeks	Dubai - 4 years	1%
UV/Cond	40 weeks	Dubai - 6½ years	5%
UV/Cond	40 weeks	S.Leone - 4 years	5%
UV/Cond	52 weeks	UK - 2 years	1%
UV/Cond	52 weeks	UK - 6½ years	0.1%
UV/Cond	52 weeks	Dubai - 2 years	5%
UV/Cond	52 weeks	Dubai - 4 years	5%
UV/Cond	52 weeks	Dubai - 6½ years	5%
Heat only	8 weeks	UK - 6½ years	5%*
Heat only	16 weeks	UK - 2 years	5%*
Heat only	16 weeks	UK - 6½ years	5%*

* Negative correlations

Table 7

Summary of significant correlations of % modulus change between natural weathering and freshly cured laboratory test specimens.

Laboratory Test		Natural Weathering		Significance Level
Conditions	Duration	Location	Duration	
UV/Cond	2 weeks	UK	1 year	5%
UV/Cond	4 weeks	UK	1 year	5%
UV/Cond	4 weeks	UK	2 years	5%
UV/Cond	8 weeks	UK	1 year	5%
UV/Cond	8 weeks	UK	2 years	5%
Heat only	2 weeks	UK	1 year	5%
Heat only	4 weeks	UK	1 year	5%
Heat only	4 weeks	UK	2 years	5%
Heat only	28 weeks	Dubai	4 years	5%*
Heat only	28 weeks	Dubai	6½ years	5%*
Heat only	28 weeks	S Leone	4 years	1%*
Heat only	40 weeks	S Leone	4 weeks	5%*

* Negative correlations

It is apparent that the freshly cured samples, table 7, tend to require only a relatively short period of artificial ageing, heat or UV and condensation, to correlate with the changes that occur after one or two years of natural weathering. This is in agreement with the results from the freshly cured sealants referred to earlier, table 3. However this is applicable to only to samples that were exposed in the UK. The correlations obtained for freshly made samples with those exposed for 4 or 6½ years in Dubai or Sierra Leone are negative. These four negative correlations are all produced after relatively long periods of exposure to heating.

CONCLUSIONS

1. Some sealants can take over twelve months before modulus values stabilize, when stored in standard conditions of $23\pm2^{\circ}$C and $50\pm5\%$ rh.

2. Changes in modulus after one or two years exposure to natural weather correlate well with changes found after short periods of artificial weathering of freshly cured sealants.

3. Changes in modulus after four or six and a half years of exposure to natural weather correlate well with changes found after longer periods of artificial weathering of well cured samples.

REFERENCES

1. Beech, J.C. and Beasley, J.L., "Evaluation of cure and durability of building sealants", Science and Technology of Building Seals, Sealants, Glazing and Waterproofing: Second Volume. ASTM STP 1200 Jerome M. Klosowski, Ed., American Society for Testing and Materials, Philadelphia, 1992.

2. Beech, J.C. and Beasley, J.L., "Further studies of cure and durability of building sealants", Science and Technology of Building Seals, Sealants, Glazing and Waterproofing: Third Volume. ASTM STP 1254 James C. Myers, Ed., American Society for Testing and Materials, Philadelphia, 1994.

3. Beech, J.C. and Beasley, J.L., "Effects of natural and artificial weathering on building sealants", Science and Technology of Building Seals, Sealants, Glazing and Waterproofing: Fourth Volume. ASTM STP 1243 D. Nicastro, Ed., American Society for Testing and Materials, Philadelphia, 1994.

4. British Standard Specification BS 5215 : 1983. One part gun grade polysulphide based sealants. BSI. London, 1983.

5. British Standard Specification BS 4254 : 1986. Two part polysulphide based sealants. BSI. London, 1986.

6. British Standard Specification BS 5889 : 1989. One part gun grade silicone based sealants. BSI, London, 1989.

3 EFFECT OF DEPTH ON FATIGUE RESISTANCE OF SEALING BEADS OF VARIOUS RECTANGULAR CROSS SECTIONS

K. TANAKA
Tokyo Institute of Technology, Yokohama, Japan
M. KOIKE
Chiba Institue of Technology, Narasino, Japan

Abstract
Fatigue resistance of rectangular sealing beads of various width and depth were experimentally investigated to support sealed joint designs, in particular, how to decide dimensions of depth of them reasonably. First, fatigue tests were carried out for a polysulfide sealant at three levels of amplitude of joint movement, $\pm 10\%$, $\pm 30\%$ and $\pm 50\%$ of initial joint width. Beads of shallow cross-section buckled like an arch at joint contraction, while deep ones bulged like a barrel. The positions where cracks appeared closely related to the compressed shape during fatigue operation; Cracks mostly appeared in the middle of bottom surface of beads when buckled, and at the corners on both exterior and bottom surfaces when bulged. It was made clear through photo-elastic analysis that the positions of crack initiation correspond with the area in beads where tensile and compressive stresses appeared repeatedly in response to joint movement. Fatigue resistance of sealing beads was not simply related to their proportions of depth(D)/width(W), but much affected by depth itself. The number of cycles to crack initiation(N) can be expressed by log N $= -A\log(D^{n}/W)+B$, where A and B are constants. Finally, additional fatigue tests were carried out to determine whether the experimental equation can be applied to other sealants.
Keywords: Sealant, Sealing bead, Dimension, Fatigue resistance, Joint movement, Photo-elasticity, Stress

1 Introduction

The matter of concern in designing sealing beads is to ensure long serviceability of them, in particular fatigue resistance to joint movement. The most important

Durability of Building Sealants. Edited by J.C. Beech and A.T. Wolf. © RILEM.
Published by E & FN Spon, 2–6 Boundary Row, London SE1 8HN, UK. ISBN 0 419 21070 9.

process in it is to decide size of beads, i.e., to decide width and depth of them.   As for width of them, the design method which is based on both joint movement and movement capability of a sealant has been widely used in our country and its validity has been proved by many applications to actual joints.

As for depth of beads, however, it is still at the empirical stage without a reasonable technical background.   The first researcher who studied the subject was Egons Tons[1]. He theoretically proved that elongation of a sealing bead depended on the shape factor which is defined as the ratio of depth(D) to width(W), D/W. For joints of ordinary width, we have been designing depth of beads considering both his study and our experiences[2].

Recently, the joints which are wider than before are often used, in particular, for wall panels for a ground floor, because they are higher than those of ordinary floors, and consequently, larger joint movements are expected.   However, we don't have any reasonable methods to design depth of beads.   It became a matter of concern again how to decide depth of beads, in particular, for wider joints of over 20mm or 30mm wide.

In this paper, we carried out fatigue tests on various sizes of sealing beads and discussed the effect of depth on fatigue resistance.

2   Specimen

A sealed joint is constructed by injecting a sealant into the space of panels and a back-up material inserted between them.   Then, a two-part polysulfide sealant, which conforms to the Japanese Standard A 5758, was injected into the space between two aluminum bars as shown in Fig.1, at the bottom of which a formed polyethylene back-up material had been installed and it was taken away just before a fatigue test for observation on bottom surface of a bead.   The sealant was cured for more than one month at room temperature.

Cross-sections of beads are 5mm,10mm,20mm or 40mm wide and 5mm, 10mm, 20mm or 40mm deep as shown in Fig.2.   For the width of 20mm, which is widely designed for actual joints in our country, specimens of 7.5mm or 15mm deep were furthermore provided to know the effect of depth of beads in detail.

3   Fatigue Test

Two specimens for each cross-section were fixed to the fatigue test equipment as shown in Fig.3, and were extended and contracted repeatedly according to locus of sine curve.   The amplitude of movement was ±30% of an initial joint width for all specimens.   For the specimens of 20mm wide, the amplitude of ±10% and ± 50% were also added to determine the effect of the deformation rate.   The period of movement was 10 seconds.   We adopted this accelerated velocity of movement to get test results in a short time, though it is rather shorter than those observed in actual joints.

Face and bottom surfaces of the specimens were periodically inspected with the naked eye directly, or through a fiber scope for bottom surface.   The numbers of

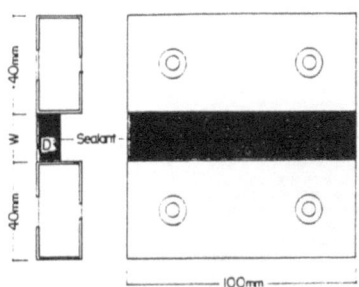

Fig.1. Test specimen for fatigue test

Fig.3. Fatigue test apparatus

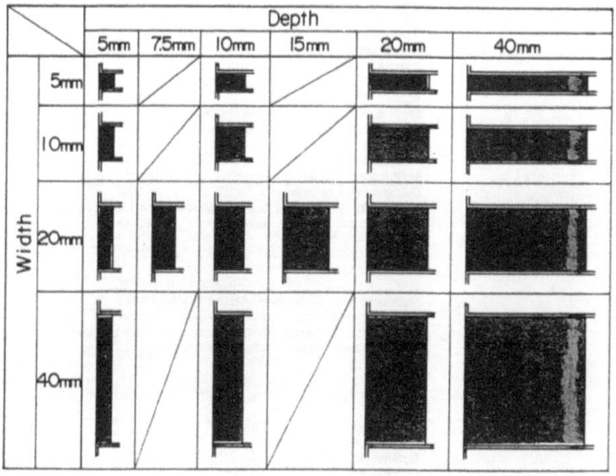

		Depth					
		5mm	7.5mm	10mm	15mm	20mm	40mm
Width	5mm						
	10mm						
	20mm						
	40mm						

Fig.2. Cross sections of sealing beads

repetition to crack initiation and rupture, and the positions of cracking were recorded.

4 Test Results

4.1 Fatigue resistance
Results of the fatigue tests are shown in Fig.4. No defects were observed in any specimens until 10,000 cycles of movement at the amplitude of ±10%. Cracks appeared on surface of specimens at the amplitude of ±30% except a few specimens of shallow cross-sections in depth. Some of the cracked specimens completely ruptured in the following repetition. At the amplitude of ±50%, cracks appeared for all specimens in early stages of tests and ruptured within 10,000 times' repetition.

4.2 Cross-sectional shape of bead at extension and contraction
Rectangular cross-sections of beads change simply to shape of a concavo-concave

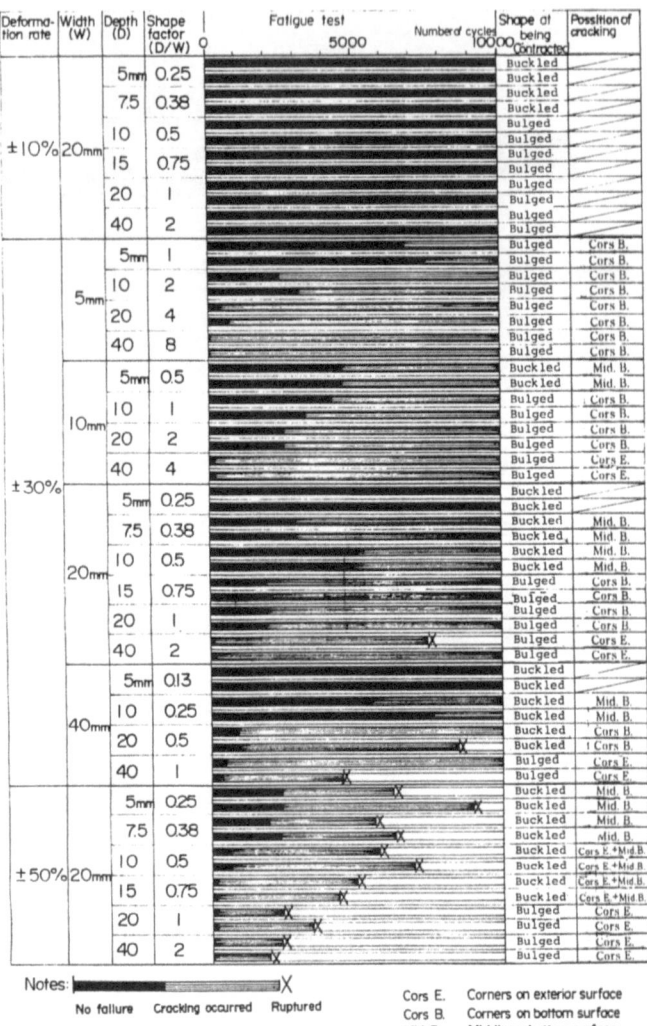

Fig.4. Results of fatigue tests

lens as the joint was widened. They, however, change to two kinds of cross-sectional shape as the joint was narrowed as shown in Fig.5, according to shape factors of beads. Shallow beads of small shape factor buckle like an arch as shown in the left of Fig.5. On the other hand, deep beads of large shape factor deform like a barrel in the right of it. It is quite natural that cross-sectional shape is also affected by contraction rate, and the boundary values of shape factor were between 0.38 and 0.5 at contraction rate of 10%, 0.5 and 0.75 at 30% contraction, and 0.75 and 1.0 at 50% contraction respectively. Beads buckle at under the boundary values and become barrel shapes above the values.

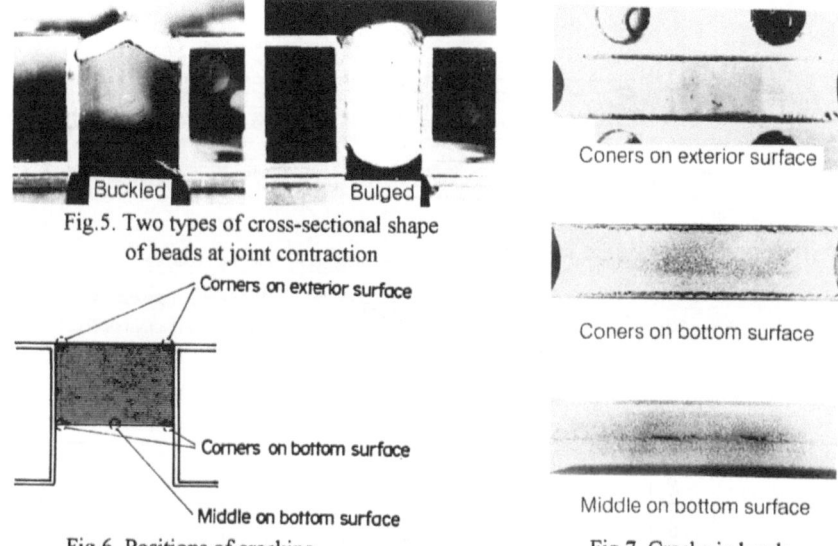

Fig.5. Two types of cross-sectional shape
of beads at joint contraction

Fig.6. Positions of cracking

Fig.7. Cracks in beads

4.3 Positions of cracking

Cracks due to fatigue were observed in the three positions on surface of sealing
beads as shown in Fig.6; (a) at the corner on exterior surface in cross-section, (b) at
the corner on bottom surface, (c) in the middle on bottom surface. Fig.7 shows
typical cracks which appeared at the above three positions. Cracks first appeared at
one of the above three positions, and they became gradually wider and deeper by
further fatigue operations. In some specimens, additional cracks newly appeared at
another positions of beads at the amplitude of ±50%.

5 Discussion

5.1 Relation between position of cracking and deformed shape of beads.

It seems that the positions of cracking in beads are related tocross-sectional shape on
being deformed, in particular, to the shape at being contracted. We can read the
close relation between them from Fig.8; cracks appear in the middle on bottom
surface of beads when they buckle, and at corners when they bulge respectively.

5.2 Relation between stress distribution and position of cracking

The above results must be related to stress distribution in beads, so we tried to
analyze surface stresses of beads induced by joint movement through a photo-elastic
technique. This is one of the methods which are suitable for analyzing stress of
large deformation materials such as rubber, and it has also an advantage that stress
concentration can be visually observed.

Commercial sealants, unfortunately, can not be used for a photo-elastic
measurement, because they are not transparent and do not produce enough fringes
for analyzing their stresses. Therefore, a translucent and optically sensitive

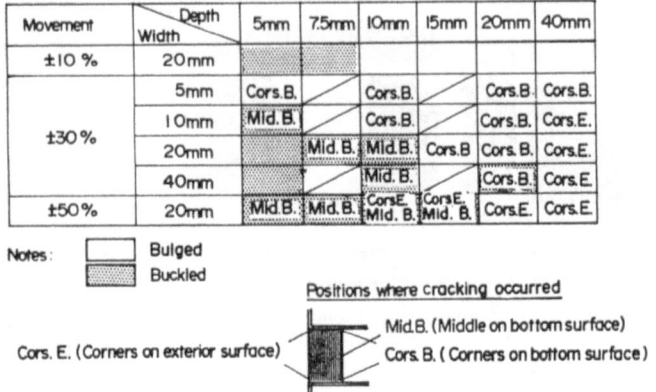

Movement	Depth / Width	5mm	7.5mm	10mm	15mm	20mm	40mm
±10 %	20mm						
±30 %	5mm	Cors.B.		Cors.B.		Cors.B.	Cors.B.
	10mm	Mid. B.		Cors.B.		Cors.B.	Cors.E.
	20mm		Mid. B.	Mid.B.	Cors.B	Cors. B.	Cors.E.
	40mm			Mid. B.		Cors.B.	Cors.E.
±50%	20mm	Mid.B.	Mid. B.	CorsE Mid. B.	CorsE Mid. B.	Cors.E.	Cors.E.

Notes: ☐ Bulged   ▦ Buckled

Positions where cracking occurred

Cors. E. (Corners on exterior surface)   Mid.B. (Middle on bottom surface)   Cors. B. (Corners on bottom surface)

Fig.8. Relation between shapes of beads at contraction and positions of cracking

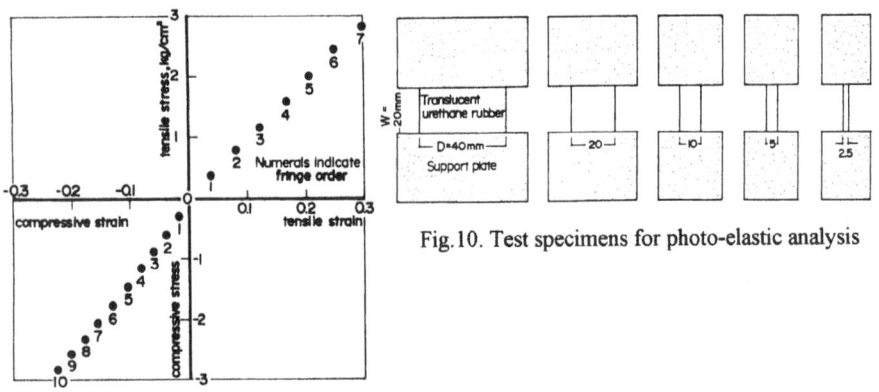

Fig.9. Stress-strain relation and fringe order
of urethane rubber sheet

Fig.10. Test specimens for photo-elastic analysis

urethane rubber sheet of 6mm thick was selected instead of them, because it has a relatively similar mechanical property to sealants among photo-elastic materials. Fig.9 shows the stress-strain relation and the fringe order of it.

Five kinds of cross-sectional model specimens of 20mm wide and 2.5mm to 40mm deep, of which shape factors are 0.125 to 2, were provided as shown in Fig.10. The specimens were inserted between two sheet glasses to prevent out-of-plane buckling at contraction and were placed in a polariscope.   They were extended and contracted to 10% and 30% of the initial joint width, and isochromatic patterns induced in the specimens by monochromatic light from the mercury lamp were observed.

Isochromatic figures and free perimeter stress distributions are shown in the upper halves and the lower halves of Fig.11 respectively.   When the specimens were extended, they deformed to concavo-concave shape and symmetric tensile stress appeared along both free perimeters.   Perimeter tensile stresses are influenced by the depth of beads, and they become slightly larger at the corners according to increase of depth, while the stresses at the middle change little.

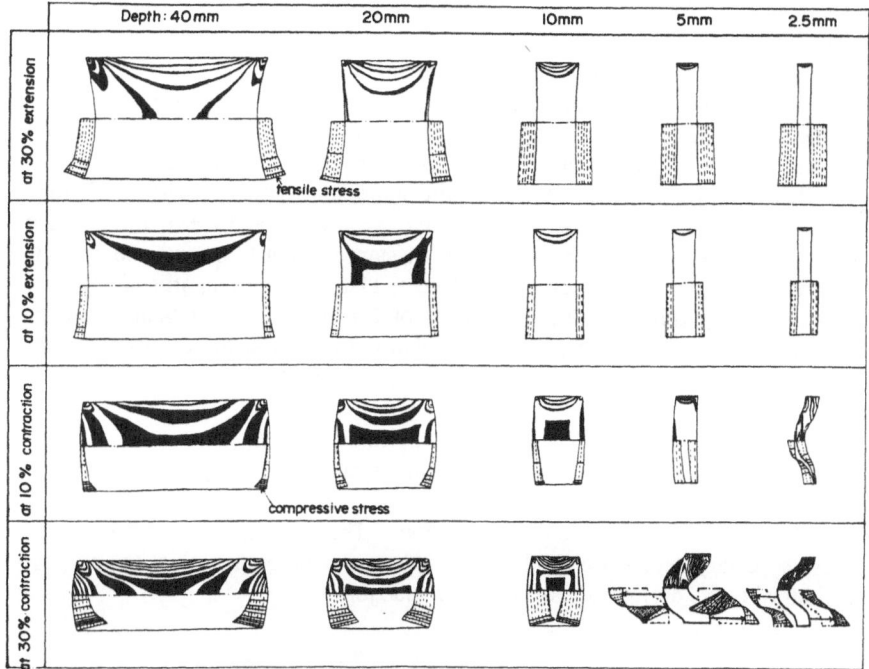

Fig.11. Isochronic figures and perimeter stresses

Two types of shape were observed when the specimens were contracted.   The specimens, of 10mm depth or deeper, were simply contracted and deformed to convexo-convex shape like a barrel, and those of 2.5mm and 5mm deep, on the other hand, buckled like an arch.   In the former, only compressive stresses were observed along the perimeter of the specimens and they gradually increased at the corners according to increase of depth.   In the latter, opposite stresses were observed on exterior and bottom surfaces of a bead.   On exterior surface of it, compressive and tensile stresses appeared at the corners and in the middle respectively, and on the bottom surface, tensile and compressive stresses conversely appeared at the corners and in the middle respectively.

Judging from the results from the photo-elastic analysis, positions where cracks occurred seem to be closely related to the area in beads where large compressive stress and large tensile stress appeared by turns during fatigue operations.   It happens at the corners of beads when specimens deform to concavo-concave shape and in the middle when they buckle like an arch at joint contraction.   This fact suggests the importance of compressive stress for estimating fatigue resistance of sealing beads.

5.3   Relation between cross-sectional size of beads and fatigue resistance

A shape factor, D(Depth)/ W(Width) is commonly used in our country as an index for deciding dimension of depth of sealing beads.   Therefore, the relation between shape factor and fatigue resistance of sealing beads is furthermore studied here.   It is shown clearly in Fig.12 that fatigue resistance of beads becomes lower according to increase of the shape factor in both tests of amplitude of $\pm 30\%$ and $\pm 50\%$. (No figure is made because no failures occurred in the test of amplitude of $\pm 10\%$.) Fatigue resistance, however, is not proportional to the shape factor.   This means that the effect of depth and the effect of width on fatigue of sealing beads are not equivalent each other. These results require more study about the effect of cross-sectional dimension of sealing beads.   Then, we tried to apply a modified shape factor, D^n/W, for estimation of fatigue resistance.   Fig.13 shows the relation between them at various values of n.   It seems that there are proportional relations between the logarithmic value of number of cycles of crack initiation and the logarithmic value of D^n/W at n=5.   The relation can be roughly expressed by the following experimental equations,

$$\log N = -0.48\log(D^5/W) + 5.5 \quad \text{(Amplitude: } \pm 30\%),$$
$$\log N = -0.48\log(D^5/W) + 4.7 \quad \text{(Amplitude: } \pm 50\%),$$

where N is number of cycles to crack initiation.   These equations mean that it is not always necessary to increase depth of a bead by the same proportion as the increase of width when we design a sealing bead for wide joints.

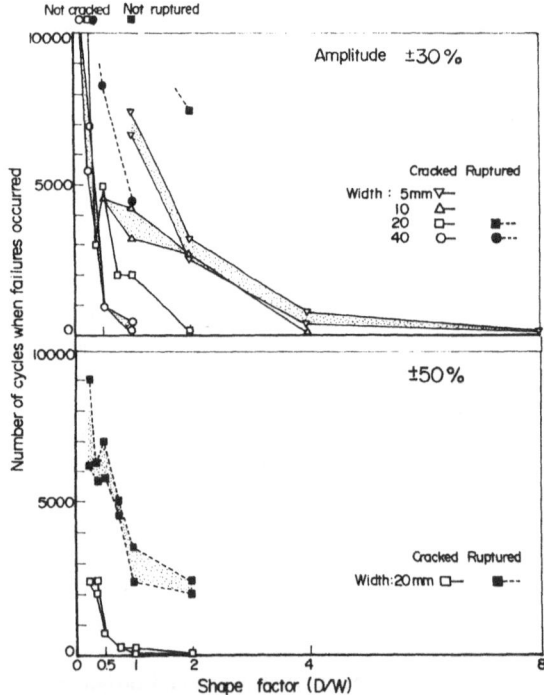

Fig.12. Relation between shape factor(D/W) and number of cycles to
crack initiation and rupture

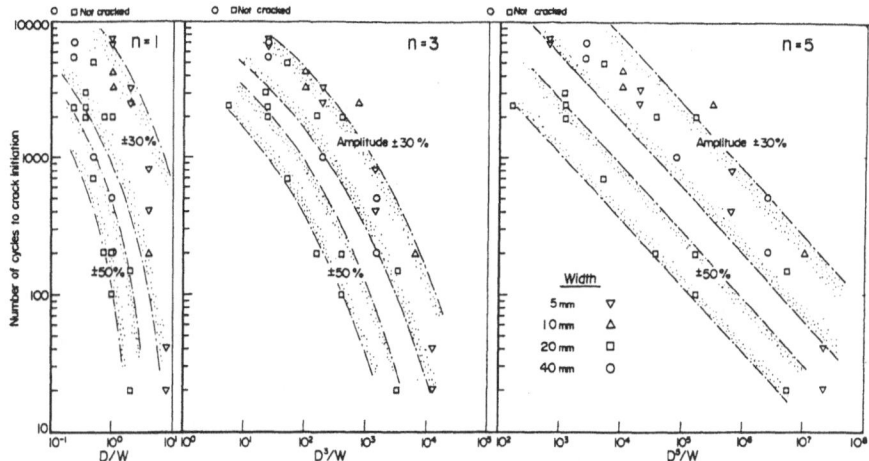

Fig.13. Relation between D^n/W at n=1,3 and 5, and number of
cycles to crack initiation for polysulfide sealant

6 Effect of cross-sectional size on fatigue resistance of other sealants

We furthermore carried out fatigue tests for other commercial sealants in our country
such as a one-part silicone sealant, a two-part silicone sealant, a modified silicone
sealant and a polyurethane sealant to examine whether the equation obtained above
can be also accepted for them.   The test method and procedure are mostly the same
as those of the previous section, except test specimens are 200mm long and one
specimen for each cross section was tested at ±30% of amplitude in this test.

As for the effect of cross-sectional size, a similar tendency to the polysulfide
sealant was observed as shown in Fig.14.   Then, we also tried to derive the
experimental equations from the results and show them in Fig.14.

7 Conclusion

Fatigue resistance of sealing beads of various rectangular cross-sections has been
studied to make clear the effect of their depth and the following conclusions were
obtained.
(1) Positions where cracks appeared by fatigue of sealants are affected by the
deformed shapes of beads, particularly to those being contracted.   Beads of deep
cross-sections deform to convex shape as a barrel and shallower ones buckle as an
arch, and most cracks were observed at the corners of beads for the former, and in
the middle for the latter.
(2) Cracking mostly start from the area of beads where severe tensile and
compressive stresses occur by turns.
(3) Fatigue resistance of a sealing bead is not simply related to proportions of
depth/width, but affected by the depth itself.   The number of cycles to crack
initiation can be roughly estimated by the following experimental equation,
$\log N = -A\log(D^n/W)+B$, where A and B are constants.

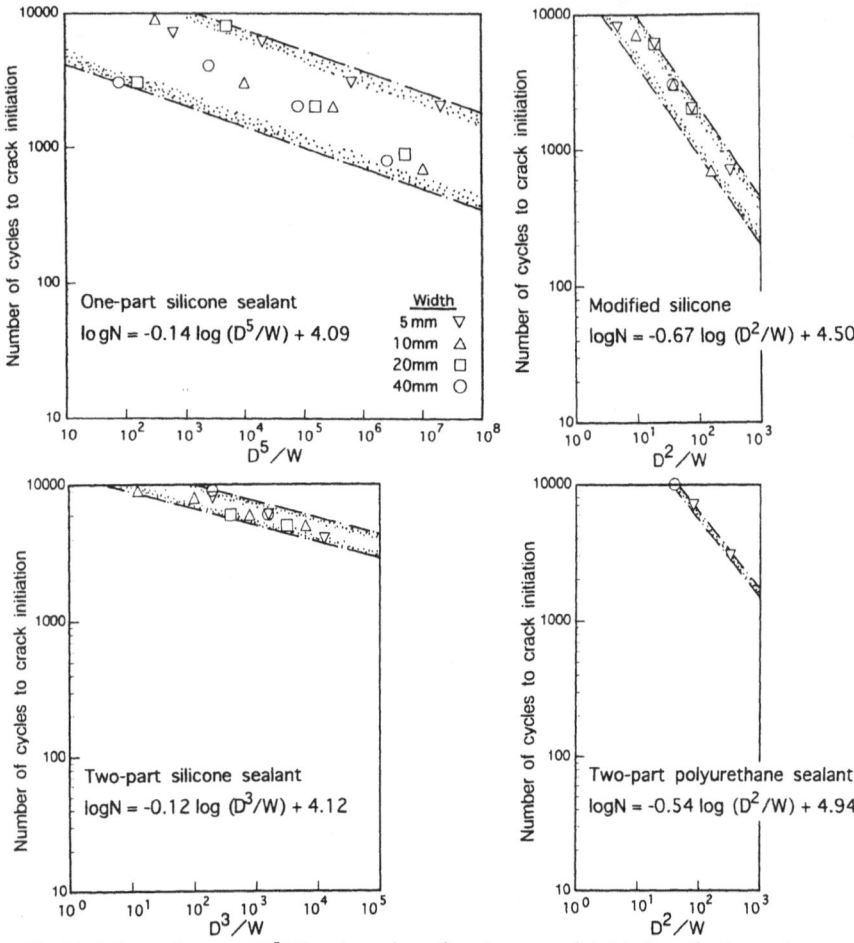

Fig. 14. Relation between D^n/W and number of cycles to crack initiation of other sealants

(4) It is advisable to reduce the shape factor accordingly as joints become wider, in particular, for joints of wider than 20mm wide.

Acknowledgments
The authors wish to thank Mr. Hanseung Oh, who is a research student, and this work was greatly assisted by his continued efforts. We also wish to express our thanks to Mr. Hideki Suzuki and Mr. Keisuke Imai for preparing the specimens.

References
1.   Egons Tons.(1959) Theoretical Approach to Design of a Road Joint Seal, *Highway Research Board* I
2.   Architectural Insitute of Japan.(1993) *Japanese Architectural Standard Specification, Water proofing and Sealing,* JASS8.

4 COMPARISON OF SHORT-TERM AGING METHODS FOR JOINT SEALANTS

M. HUGENER and S. HEAN
EMPA Swiss Federal Laboratories for Materials Testing and Research,
Section of Road Engineering/Sealing Components,
Dübendorf, Switzerland

Abstract
In this article the short-term aging of bituminous sealing compounds for bridge deck joints is described with respect to both laboratory tests and field performance. Four different products containing elastomers as a modifying agent were applied under controlled conditions to four test fields in Switzerland. Constructions of the plug joints consisted of three layers of sealants, which were placed successively and reinforced with aggregates. From each layer a sample was collected to study the aging during construction.

In addition, different laboratory aging methods were applied to the bituminous joint sealing compounds in the state of delivery: Rotating flask at 165°C (DIN 52016 [1]) and 180°C, rolling thin film oven test (RTFOT) at 180°C and storage in the oven at 70°C for 14 days (SN 671'904 [2]) and 80°C for 7 days.

Laboratory aged and non-aged samples, as well as samples from the field were characterised by penetration, ring and ball softening point and thermogravimetric analysis. The state of polymer in the sealants was investigated using gel permeation chromatography (GPC). The short-term aging methods were compared using the results from the different characterising methods described above. The states of the joint sealants from the field objects were compared and correlated.

The change of physical parameters during installation of the bridge desk expansion joints corresponds approximately to the change found for laboratory thermal oxidative aged samples. According to GPC the degree of polymer degradation in the polymer bitumen joint sealants during installation corresponds approximately to thermal aged laboratory samples.
Keywords: bituminous joint sealants, bridge deck joints, aging methods, gel permeation chromatography.

Durability of Building Sealants. Edited by J.C. Beech and A.T. Wolf. © RILEM.
Published by E & FN Spon, 2–6 Boundary Row, London SE1 8HN, UK. ISBN 0 419 21070 9.

1 Introduction

The last few years saw an increase in the number of bridge deck expansion joints with polymer modified bitumen joint sealants, as opposed to the classical system with comb and tooth joints. One reason is that the new system offers not only reduced noise and improved driving comfort, but also increased security against penetration of solid and liquid substances (primarily salt water) into the construction. In addition, both installation and maintenance of the joints are simpler and cheaper than with conventional systems.

The increased use of polymer modified bitumen joint sealants for bridge deck expansion joints in Switzerland and elsewhere, and the special needs of quality assurance require the development of suitable new tests for the sealants with emphasis on the aging aspects. However, it is remarkable that to date, only little is known about the long-term behaviour and the effects of short-term thermal aging, caused by improper installation of polymer modified bitumen joint sealants.

For these reasons, research on wide bridge deck expansion joints was initiated at EMPA to investigate various laboratory aging methods and to compare them with actual short-term and long-term aging of expansion joints on weather exposed field objects [3]. Different test methods were examined for their suitability to determine and characterise the aging behaviour of joint sealants for bridge deck expansion joints. Because it was intended to monitor the objects over long periods of time and in order to minimize damage to the expansion joints due to sampling, it was important to select small sample size test methods.

This report presents first results of thermal short-term aging, that simulates installation of broad bridge deck expansion joints and various laboratory aging methods. It gives also a discussion of possible test procedures for the characterization of aging. A study on the long-term behaviour of the objects monitored will follow as soon as sufficient data are available.

2 Description of the bridge deck expansion joints

Researchers from EMPA monitored the installation of the new polymer modified bitumen joint sealants and took samples on four bridge deck expansion joints during renovation.

On the newly installed expansion joints, the expansion gap of about 20 mm was caulked with a rubber profile and then covered with a metal flashing (Fig. 1). Above this, a three to four-layer course of aggregate stabilized joint sealant was installed over a width of 400 mm and to a depth of 60 mm. Three of the four joint sealants JS1 .. JS4 were supplied by the same manufacturer but had different compositions.

The surface of the bridge deck expansion joint was cleaned over the width of the joint before a 5 mm thick layer of hot joint sealant was applied. Above this was three layers of aggregate with joint sealant, each approximately 20 mm thick. For each layer the hot aggregate was tipped and filled with hot joint sealant. The application temperature recommended by the manufacturers was 170-190°C for the joint sealant and 150°C for the aggregate. Joint sealants JS1, JS2 and JS4 were heated in a temperature controlled electrically heated drum, whereas for JS3, a gas-heated oven without temperature control was used. The aggregate was heated to the required temperature with a flame.

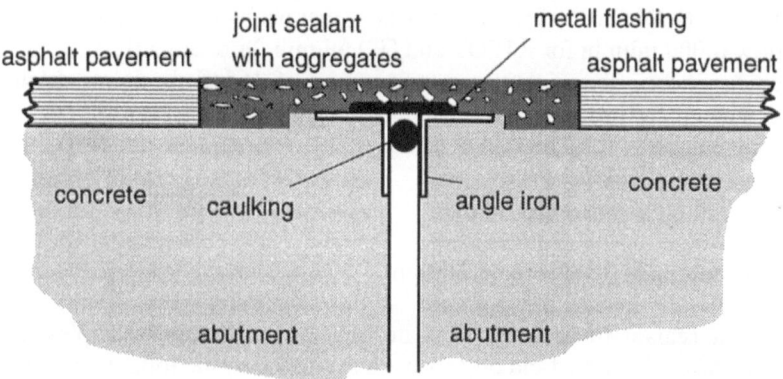

Fig. 1. Section through an asphaltic plug joint.

The temperatures of aggregate and joint sealant were measured and recorded by EMPA both in the mixer and during application. From each layer a sample of the joint sealant was poured directly into prepared aluminium moulds.

3 Aging methods in the laboratory

In order to compare the degree of joint sealant aging during installation with aging in the laboratory, the samples of sealants taken in the original state were laboratory aged in two ways:

Thermal aging methods:
- Conditioning in the oven at 70°C for 14 days
- Conditioning in the oven at 80°C for 7 days

Thermal oxidative aging methods:
- Rolling thin film oven test (RTFOT) at 180°C, 75 min.
- Rotating flask aging at 165°C, 150 min.
- Rotating flask aging at 180°C 150 min.

3.1 Thermal aging methods
According to Swiss Standard SN 671 904 [2], oven conditioning of joint sealants for 7 days at 70°C is required, but earlier investigations at EMPA

showed that this period is too short. For this reason, oven storage was extended to 14 days. As an alternative to longer conditioning time, raising the temperature by 10°C was also considered so conditioning was performed for 7 days at 80°C, which was supposed to produce a similar effect.

3.2 Thermal oxidative aging methods

Laboratory aging of the bitumen binder in road construction is frequently carried out with the rolling thin film oven test (SN 671 752c [4]) or the rotating flask aging (DIN 52016 [1]). In both cases, thermal aging occurs under constant air flow (4000 ml/min for RTFOT and 500 ml/min for rotating flask).

Since experience shows that at a temperature of 165°C the joint sealants are often incompletely molten and hence inhomogeneous, RTFOT was performed at a temperature of 180°C instead of 165°C. The rotating flask test was conducted both at 165 and 180°C.

4. Test methods for characterization

The joint sealant in the original state, the laboratory aged samples and the samples taken from the field objects were examined by the following tests:

- Penetration needles (25°C)
- Penetration cones (25°C)
- R&B softening point
- Thermogravimetric analysis (TGA)
- Gel permeation chromatography (GPC)

4.1 Penetration needles and cones, R&B softening point

In Switzerland the penetration of joint sealants is determined with a needle, whereas in most other countries a cone is used. For this reason, both tests were carried out to see whether a correlation between the two methods exists. In the R&B softening point test, rings with shoulders were used.

4.2 Thermogravimetric analysis (TGA)

In thermogravimetric analysis [5] a sample of approximately 10 mg was heated from 25 to 750°C and the change of mass, due to evaporation and degradation, was measured as a function of the temperature.

For this investigation, the following temperature program was used:

- Room temperature to 180°C: heating rate 10°C/min (heating up phase)
- 60 min isothermal at 180°C
- 180 to 750°C: heating rate 20°C/min

The analysis gives information on the fraction of volatile substances and of polymers, bitumen and fillers, which degrade and evaporate at different temperatures. In Fig. 2 this is shown for the joint sealant JS2 in original state:

* Range A (from 25 to 180°C): The joint sealant becomes liquid and volatile compounds such as oils, solvents and also water evaporate.
* Range B (from 180 to 390°C): The bitumen fractions start to degrade.
* Range C (from 390 to 600°C): In this range, the various components of bitumen and also the polymer degrade.
* At 600°C: at this temperature only ash remains
* Range D (from 600 to 750°C): Binder ash consisting of calcium carbonate is converted to calcium oxide.

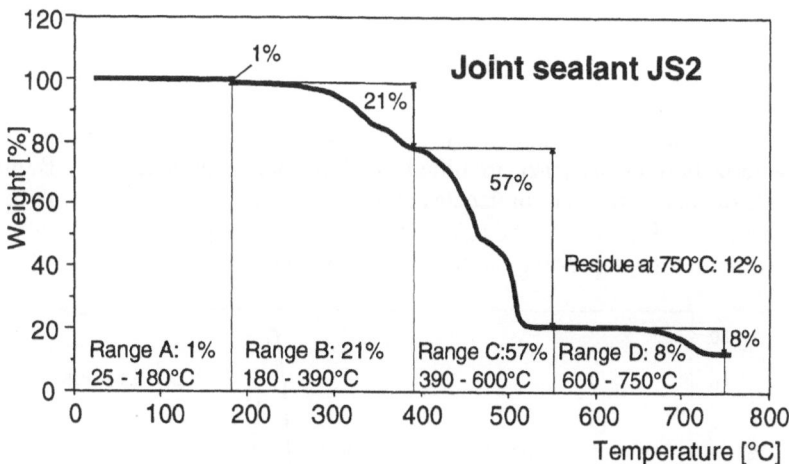

Fig. 2. TGA-Diagram of joint sealant JS2 in original state.

4.3 Gel permeation chromatography (GPC)

Gel permeation chromatography is a chemical method. The joint sealant is dissolved in a suitable solvent and the individual components separated according to their size, which is a function of the molecular weight. The dissolved sample is pumped through a column filled with microscopic spheres (10 μm diameter), whose surface is covered with countless pores of defined size. Small molecules can diffuse into these pores, while large molecules do not fit inside and flow by in the solvent without delay. Thus, the large molecules emerge first from the column, followed by the remaining molecules in order of decreasing size.

The substances are detected with an RI detector (refractive index measurement) or an UV detector (measurement of the absorption of UV light). The intensity of the signal is proportional to the concentration and is recorded in a chromatogram as a function of the analysis time. Thus the individual substances are recognized as peaks. In the chromatogram, the size of the molecules decreases logarithmically with increasing analysis time.

The conditions for the chromatographic analysis are:

Column:	1 column 60 cm, PLgel 5µ MIXED-C (Polymer Laboratories)
Solvent:	Tetrahydrofuran (THF)
Injection volume:	20 µl
Oven temperature:	30°C
Pump temperature:	30°C
Injector Temperature:	30°C
Flow rate:	1.0 ml/min
Pressure:	40 bar
Run time:	40 min
UV-wavelength:	from 200 to 450 nm

Gel permeation chromatography is suitable for the separation and characterization of polymer modified binders and joint sealants. Bitumen consists of relatively small molecules and is recognizable as a large peak in the chromatogram (Fig. 3). To the left of it, several smaller peaks appear, with size and retention time depending on the type of polymer.

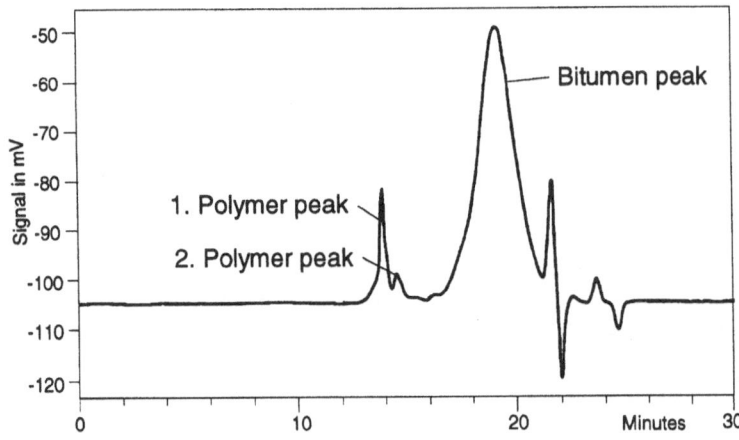

Fig. 3. Chromatogram of joint sealant JS2 in original state.

During thermal aging of joint sealants the polymer deteriorates more strongly than the bitumen. The long polymer chains are fractured and shorter polymers with lower molecular mass are formed. This is seen in the chromatogram as a reduction of the polymer peaksize and new, broader peaks from the degraded polymer. A comparison of the chromatograms from aged samples with those in the original state shows how the polymer in the joint sealant has changed

(Fig. 4). The height or area of the first polymer peak is proportional to the concentration of the non-degraded polymer portion and can be used for quantification.

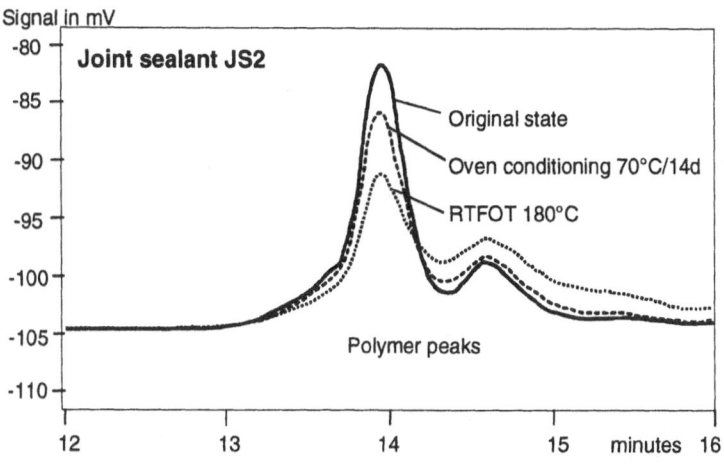

Fig. 4. Comparison of different aging methods for joint sealant JS2.

5 Results

5.1 Penetration
The diagram of the penetration cone results (Fig. 5) shows clearly that JS1 and JS2 behave differently from the other two joint sealants. Whereas JS1 and JS2 display a reduction of penetration on aging by rotating flask at 180°C, the

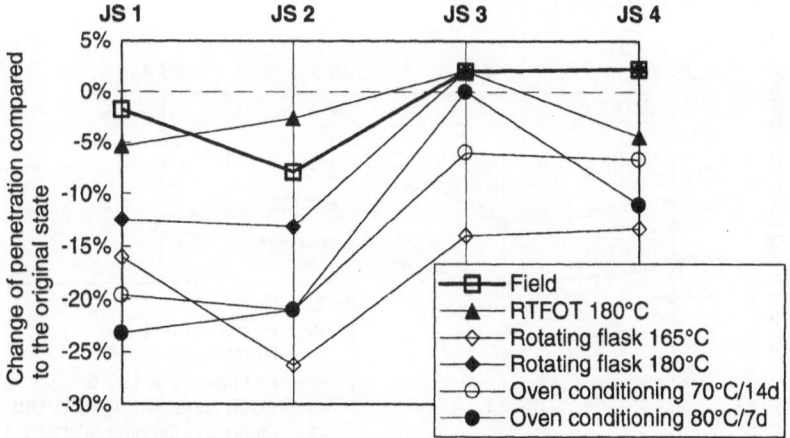

Fig. 5. Penetration with cone of the joint sealants JS1..JS4.

values for JS3 and JS4 increase slightly. Both types of conditioning in the oven lead to similar results, except for JS3.

The field samples are very different from the laboratory aged samples and agree best with the RTFOT samples.

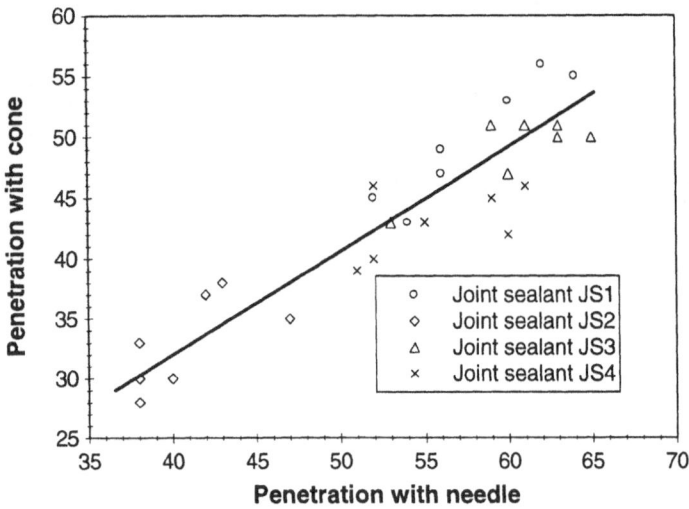

Fig. 6. Correlation between the two penetration methods.

The results of the investigation show that for penetration measurement, both methods (needles or cones) may be used. Fig. 6 displays a correlation between penetration needles and cones ($R^2 = 0.85$).

5.2 Ring and ball softening point

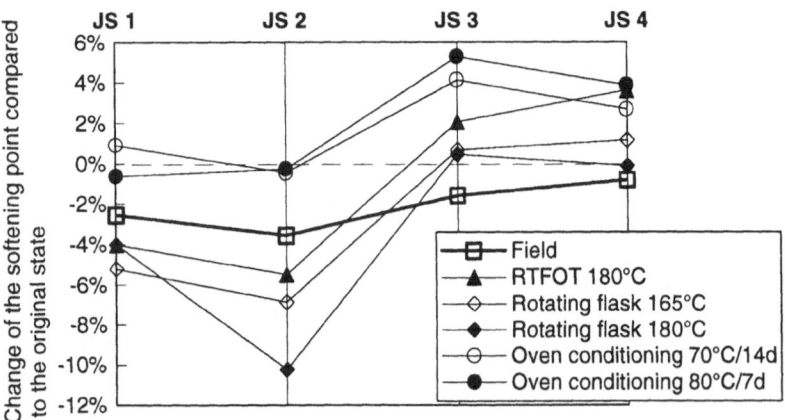

Fig. 7. Comparison of ring and ball softening points.

Again the laboratory aging methods show very different influences on the sealants JS1 and JS2 compared to JS3 and JS4 (Fig. 7). Whereas the softening point for JS1 and JS2 drops clearly in some cases, the value for JS3 and JS4 is slightly larger in general. On the other hand, a clear difference is observed between the storage in the oven and the other thermal oxidative aging methods. A strong reduction in softening point is observed with JS1 and JS2 in the RTFOT and rotating flask aging, but conditioning in the oven shows practically no influence. This could be caused by atmospheric oxygen, which has a smaller effect on conditioning in the oven.

5.3 Thermogravimetric analysis (TGA)
The investigation of four joint sealants by TGA showed that there were only small differences between field, laboratory aged, and original state sealants (Fig. 2). These differences were in the same range as the uncertainty of the thermogravimetric analysis method. Therefore, the evaluation of the results was considered unreasonable.

5.4 Gel permeation chromatography
As shown in Fig. 8 the height of the first peaks with the original state as reference, changes considerably depending on the method used for aging. Apart from JS2, the polymer is most strongly degraded by RTFOT at 180°C. However, the difference from rotating flask aging at 180°C is often not very large. Rotating flask aging at 165°C shows very different results, probably due to inhomogeneity of the only partially molten sealant.

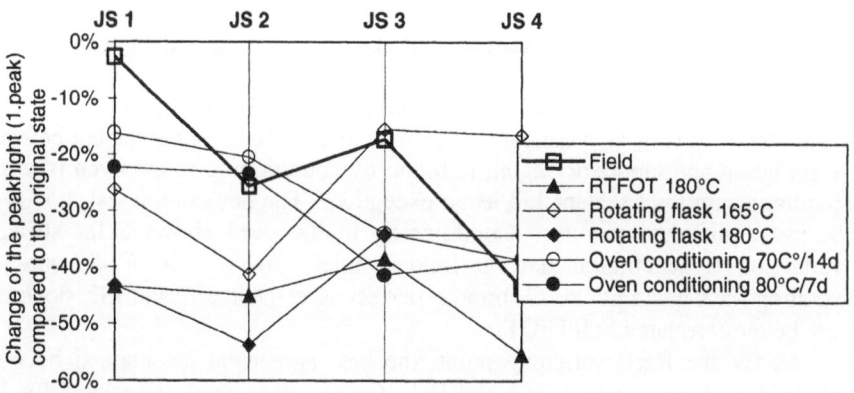

Fig. 8. Change of the polymer content determined by gel permeation chromatography.

The two conditioning types in the oven, 14 days at 70°C and 7 days at 80°C, yield almost the same results. With the exception of JS3, they show the best agreement with the samples from the field. The polymer degradation is generally somewhat larger at 80°C.

Joint sealants JS1 and JS2 present a large difference between conditioning in the oven and the oxidative aging methods. This could be an indication that certain joint sealants react more sensitively to oxygen than others.

6 Conclusions

The results of the double sampling from the same location show qualitatively good agreement and some important trends can be recognized and will be useful for future work. However, because of the relatively few bridge deck expansion joints available for this investigation, a statistical evaluation was not possible.

It is planned to expand this investigation and to investigate further bridge deck expansion joints with joint sealants from other manufacturers, and to monitor them over a longer period of time.

6.1 Comparison of laboratory aging methods

The results of the investigation show that the values obtained from conditioning in the oven for 14 days at 70°C and for 7 days at 80°C (thermal loading) are in good agreement for all test methods. It is therefore possible to reduce conditioning time by raising the temperature and to find an optimum without the result changing due to thermal aging.

Conditioning in the oven (thermal loading) can lead to results completely different from those obtained in RTFOT or rotating flask (thermal and oxidative loading), which is shown above all with JS1 and JS2. The reason for the different behaviour of these two joint sealants presumably lies in the increased sensitivity to atmospheric oxygen, since both in RTFOT and rotating flask air flow is continuous. Another reason might be the increased loading temperature.

6.2 Aging in the field compared with laboratory aging methods

From the results of gel permeation chromatography, the best agreement between installation and laboratory aging is found by conditioning in the oven (thermal loading) with joint sealant JS3 as an exception. The penetration test, however, shows a different picture. Conditioning in the oven shows a far stronger reduction of penetration than the field samples. In the case of JS3 and JS4 rotating flask aging at 180°C shows a perfect agreement, JS1 and JS2, however, are better correlated to RTFOT.

As for the R&B softening point, the best agreement is obtained between RTFOT and joint sealants JS1 and JS2. On the other hand, the results for JS3 and JS4 agree most closely with those from rotating flask aging at 180°C. From this it follows that the composition of the joint sealants plays a more important role on aging than expected.

6.3 Suitability of the test methods to characterize aging of bridge deck expansion joint sealants

The investigation shows no connection between the results of the classical tests such as penetration or R&B softening point and the polymer content, as measured by gel permeation chromatography. However, this was already observed with polymer modified binders for pavement construction, where the polymer content is generally lower. It is known from the literature [6] that differing polymer contents in the binder cause a change of penetration and softening point which is not proportional to the polymer concentration.

The question of how much the polymer content affects the overall properties of the joint sealant remains open. An important factor in the behaviour of polymer modified bitumen compounds is the degree of cross-linking of polymer and bitumen. This depends on the bitumen type, the polymer mixture, the additives and on the mixing process itself. This can exert at least as large an influence upon the overall properties of the joint sealants. The influence of temperature can lead to a change in the inner structure of the joint sealant without causing a substantial degradation of the polymer.

7 References

1. Deutsche Norm. (1988) *Thermische Beanspruchung im rotierenden Kolben und Bestimmung der Gewichtsänderung durch thermische Beanspruchung.* DIN52016 ,
2. Schweizer Norm. (1975) *Fugenfüllmaterialien.* SN671904.
3. Hean, S., Hugener, M. (1995) Aging behaviour of polymer bitumen joint sealants during installation, *Research Report EMPA No. 146002,* in preparation.
4. Schweizer Norm. (1993) *Bituminöse Bindemittel. Masseänderung bei 163°C, 75 min.* SN671752c.
5. Kopsch, H. (1989) *Erdöl und Kohle - Erdgas - Petrochemie vereinigt mit Brennstoff-Chemie,* Vol. 42, No. 12. p. 502.
6. King, G. N., King, H. W., Harders, O., Arand, W., Planche, P.-P. (1993) *Proceedings of the Association of Asphalt Paving Technologists,* Vol. 62, pp. 1-22.

5 THE PREDICTION OF LONG-TERM SEALANT PERFORMANCE FROM DYNAMIC ACCELERATED WEATHERING

S.A. HURLEY
Taywood Engineering Limited, Southall, UK

Abstract
This paper provides an overview of some aspects of a 4-year research project partly funded by the Commission of The European Communities under the BRITE-EURAM initiative and involving partners in France, Italy and the U.K. The main aim of the project is to develop an improved test methodology for assessing the long-term performance of sealants in different climatic areas. The objectives and background of the project are discussed and an account is given of the weathering/cyclic movement regimes employed during the work.
Keywords: Accelerated and natural weathering, cyclic movement, long term performance.

1 Introduction

This paper presents a brief overview of some aspects of a 4-year research project concerned with the durability of building sealants. The work is funded by the Commission of The European Communities under the BRITE-EURAM initiative and the following active participants:

- Taywood Engineering Limited, U.K. (lead partner)
- British Rail Research/Scientific Services, U.K.
- Centre National d'Evaluation de Photoprotection, France.
- Centre Scientifique et Technique du Batiment, France.
- Enichem Synthesis, Italy.
- Rhone-Poulenc Chimie, France.

Durability of Building Sealants. Edited by J.C. Beech and A.T. Wolf. © RILEM.
Published by E & FN Spon, 2–6 Boundary Row, London SE1 8HN, UK. ISBN 0 419 21070 9.

A considerable number of specimens, exposed to a wide range of conditions, have been analysed during this project using an extensive number of techniques, both physical and chemical. However, the detailed evaluation of many results is incomplete and the correlation of the output from different partners and tasks is still in progress. This latter area will be particularly important in determining the more general benefits of the work.

For these reasons, and also because certain results are still confidential, this paper is mainly focused on the approach adopted. Nevertheless, it is hoped that even an interim and limited discussion will provide a useful contribution to this Seminar on sealant durability.

2   Objectives and background

As many sealants are required to perform satisfactorily for at least 10-20 years under a variety of climatic conditions, test methods used to predict service life must be discriminating and reliable while also being suitable for routine use over relatively short time-scales. Two main approaches may be summarised as follows:

- Empirical methods where the absence of failure under standard test conditions is equated to different service requirements with the assistance of prior knowledge and experience, usually with an appropriate safety factor.
- More fundamental methods which seek to understand the mechanisms of degradation and ultimate failure, particularly the rate controlling steps, and to quantitatively define those factors which critically influence these processes.

While each methodology has advantages and disadvantages, the overall objective of this project was to further develop the second approach as it was considered to have the following potential benefits:

- It could assist in the formulation of new sealants by identifying individual components, or specific chemical linkages, which are particularly vulnerable in service.
- Any significant chemical changes in the sealant/substrate system which precede visible degradation (or easily detected variations in mechanical/adhesive properties) could provide a sensitive bench-mark for the assessment of performance. In principle, similar analysis could also allow critical areas of the sealant to be investigated, viz, the exposed surface, the interface and the zones of high stress concentration (using the interior bulk as a "control").
- Sensitive analysis, applicable to very small specimens of irregular shape, would have particular benefits for the assessment of in-service sealants thus, potentially, assisting the planning of maintenance
- Fundamental analysis can assist in the validation of more empirical methods.
- The same approach can be adapted to the investigation of other polymer-based materials used in construction.

It was recognised from the outset that a number of factors combine to increase the difficulty of research on sealant performance - the complexity of formulations, substrate variations, cohesive and adhesive integrity, the synergy between weathering and cyclic stress/strain and the effects of stress relaxation; in addition, of course, to the general complexities of weathering and polymer degradation. However, while representing a seemingly minor element in the building envelope, sealed joints are responsible for a high proportion of maintenance expenditure [1] and, therefore, warrant research investment.

A number of more specific objectives were defined to assist formulating answers to the following questions:

- Could changes within high performance sealants be induced and detected after relatively short exposure periods involving only a limited degree of acceleration?
- Were any such changes: modulated by variations in the test regime?
  significant with respect to long-term behaviour in a natural environment?
  able to provide a more general basis for the prediction of probable service-life?

These objectives were as follows:

- The development of accelerated weathering programmes which simulate three climates (temperate, desert and sub-tropical; a sub-arctic climate was also included in the initial proposal).
- The design and manufacture of simple test rigs which would enable cyclic movement to be imposed during both accelerated and natural weathering.
- The use of a wide variety of analytical techniques in order to determine which provide the most useful information for assessing durability.
- The analysis of sealants subjected to widely varying conditions, including service on real structures, in order to determine the characteristic mechanisms of degradation and to mathematically model the overall performance.

3   Sealant and substrate types

Three common substrates have been used - anodised aluminium, float-glass and pre-cast concrete with a maximum aggregate size of 10mm. The concrete coupons were sawn from larger panels, leaving an "as-cast" face which was lightly grit-blasted to provide, as far as possible, a realistic and consistent adherend surface.

Bonded specimens with a "flush-surface" were used (Type I in BS3712) in preference to the recessed design (Type 3 in BS3712) to prevent shadowing effects during exposure.

The dimensions of the sealant beads were 50mm (length) x 12mm (width) x 6mm (depth), the latter being governed by the need to extend and compress a number of specimens using relatively light-weight test rigs.

XPS analysis, used to characterise the substrate surfaces, showed that traces of cleaning solvents could be retained firmly and, also, a significant level of tin on one side of the glass. The aluminium and glass substrates were also characterised by measurement of static contact angles using standard procedures.

Six different sealants have been investigated: 3 silicones (acidic/basic/neutral cure types), 2 single-pack/moisture-curing polyurethanes (possessing aliphatic and aromatic back-bone structures respectively) and a manganese dioxide cured polysulphide (a gun-grade with a polymer content of 35% by weight). Primers were used with the polysulphide and the polyurethanes. All specimens were cured under normal ambient conditions for at least 4 weeks (and usually much longer) prior to exposure.

The following types of specimen were also utilised:

- Directly cast free films with a thickness of 2-3mm which were exposed in an unstressed state.
- Low molecular weight model compounds and linear elastomers used to investigate the stability of various linkages in the polyurethanes.
- Simplified sealant formulations in which various constituents were omitted.
- Sealant samples retrieved from various structures.

4   Accelerated and natural weathering

4.1 Exposure regimes

The sealants and many of the associated specimens were subjected to specific thermal, hydrolytic and photo-ageing regimes as part of the investigations concerned with the mechanisms of degradation. The bonded and free film specimens were exposed to natural weathering in London, Singapore and Phoenix, Arizona. Artificial weathering was also carried out using regimes designed to both simulate and accelerate the climatic conditions found in these zones i.e., temperate, sub-tropical and desert.

These regimes were developed from an examination of climatic data broadly following the principle described by Scott [2]. Thus an attempt was made to keep a realistic balance between different parameters while avoiding extreme and potentially anomalous conditions. Initially, more complex programmes were derived in which allowance was made for seasonal variations. However, the capabilities of the accelerated weathering equipment which was available to the partners varied widely. Consequently, the more simplified programmes summarised in Table 1 were utilised.

In all three cases, a xenon-arc source of ultra-violet light was used, thus providing a close reproduction of the spectral distribution of sun-light.

Other considerations, and particularly more extended experience, could well suggest that some modification of these programmes is desirable. However, further refinement was not investigated during this work.

4.2 Natural weathering - cyclic movement

It was essential to subject bonded specimens to cyclic movement in order to investigate

Table 1   Accelerated weathering regimes

Location	Chamber temperature (°C)	Relative humidity (%)	Continuous irradiance 340/300-400 nm (Wm^{-2})	Rain	
				On (Minutes)	Off (Minutes)
Singapore	45	85	0.55/58	8	12
Arizona	55	30	0.55/50	None	
London	35	55	0.40/58	3	17

the synergy between weathering and stress/strain. It was also desirable that the cyclic movement regimes provided a reasonable approximation to typical service conditions in the different climatic zones.

This objective was achieved most closely in the case of natural weathering by use of the test rig shown in Figures 1 and 2. This device is based upon the principle used earlier by Karpati [3], i.e., strain is induced by the use of materials possessing a different linear coefficient of thermal expansion. In order to achieve the required movement with a compact rig (and the anticipated variations in ambient temperature) a mild steel/plastic combination was employed, following the approach used by Beech [4].

Only one steel/plastic combination was used during this work giving a single amplitude per unit temperature change. However, as shown in Figures 3 and 4, further lengths of plastic could easily be incorporated. A lower level of movement could then be imposed upon additional specimens.

The steel/plastic "drive mechanism" of the rig was covered, thus restricting the response to changes in air temperature and eliminating rapid (and possibly unrealistic) fluctuations due to direct but intermittent sun-light. The resources of the project limited the deployment of this rig to one site (Southall, near London) and a typical response at this location is illustrated in Figure 5.

Additional natural weathering with cyclic movement is being carried out at all three exposure sites using manually operated devices which were principally designed for use under accelerated weathering. Obviously, with this type of test rig, shown in Figures 6 and 7, strain can only be varied in a step-wise manner. The movement cycles employed are summarised in Table 2.

4.3 Accelerated weathering - cyclic movement
An initial phase of accelerated weathering involved the exposure of unstressed free films and bonded specimens held at an elongation of 30%; continual weathering periods ranging from 500 to 4000 hours were used. Various control specimens were also included. These tests were used to investigate chemical and physical changes induced by weathering and a constant relatively high tensile strain.

This work was followed by exposure to regimes which utilised intermittent accelerated weathering and step-wise cyclic movement, as summarised in Figures 8-10.

These cycles were derived by assuming that the sealants were applied (or, more correctly, cured) when the joint was at a mean width. Changes were then imposed to

Fig. 1. Cyclic movement test rig (climate driven): during construction.

Fig. 2. Cyclic movement test rig (climate driven): in operation.

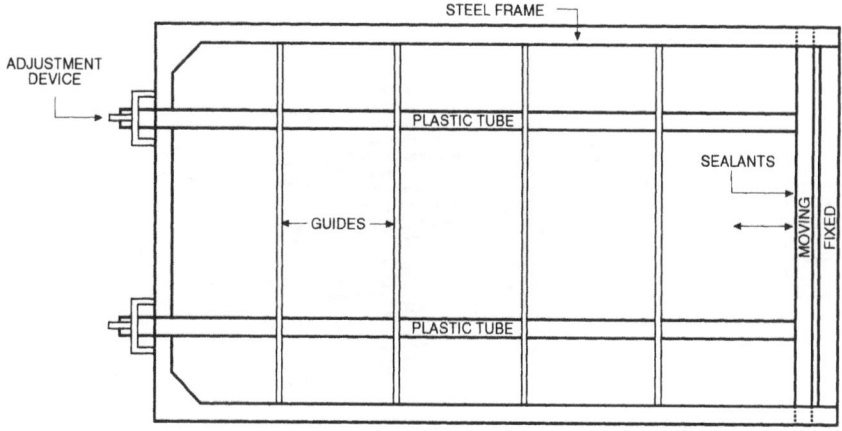

Fig. 3. Cyclic movement test rig (climate driven) : one strain level.

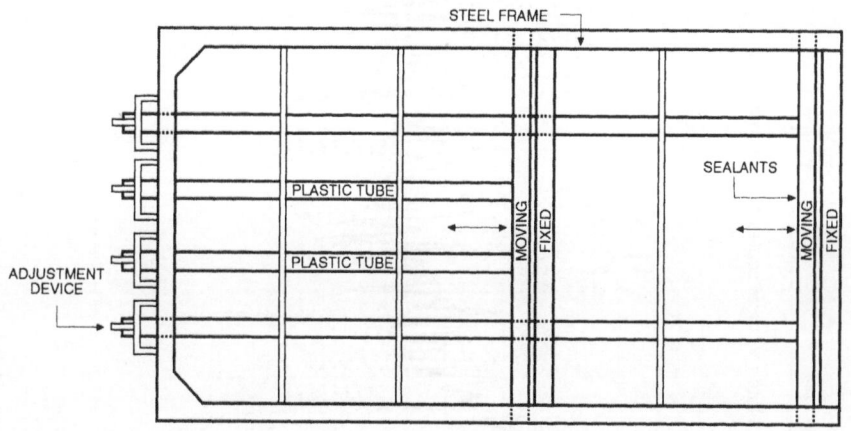

Fig. 4. Cyclic movement test rig (climate driven) : two strain levels.

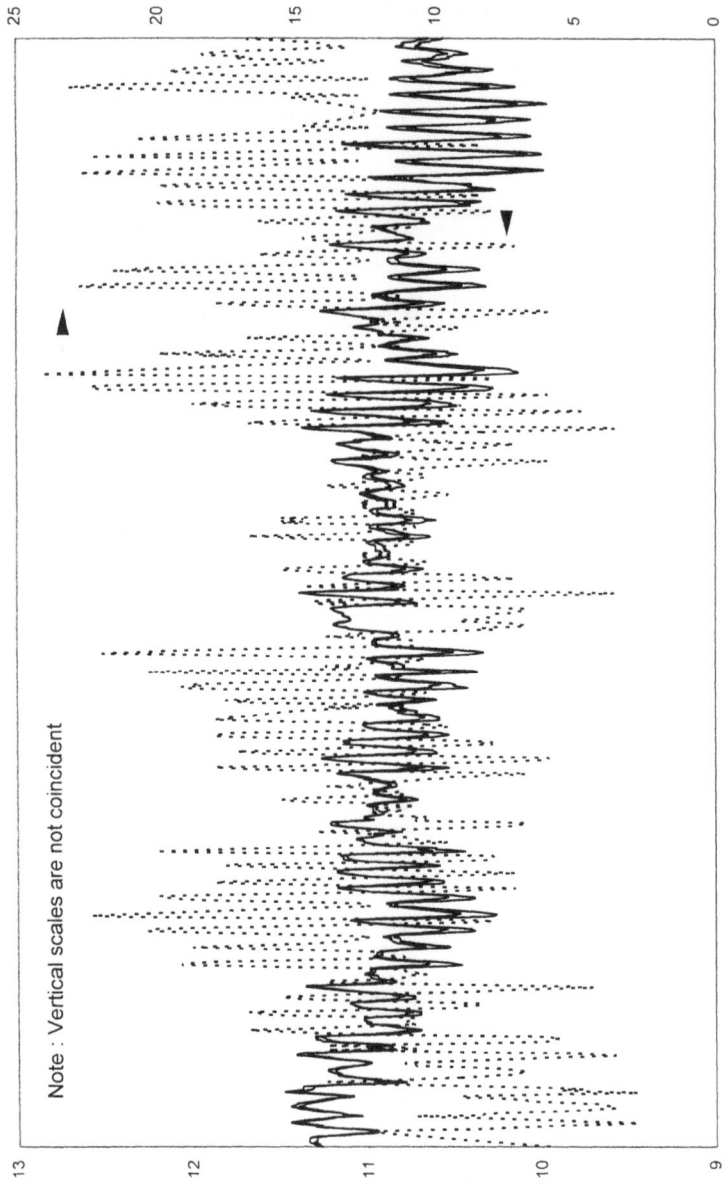

Fig.5. Climate induced movement.

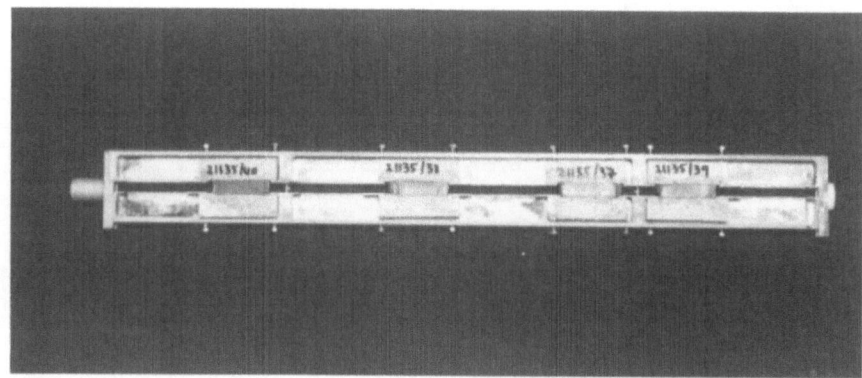

Fig. 6. Cyclic movement test rig (manually adjusted).

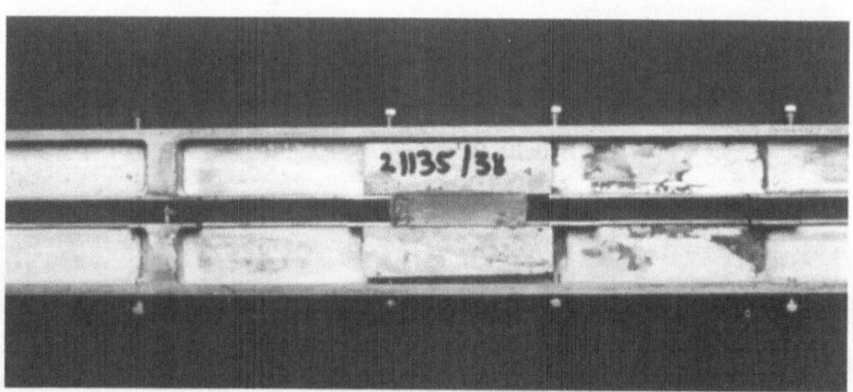

Fig. 7. Cyclic movement test rig (manually adjusted).

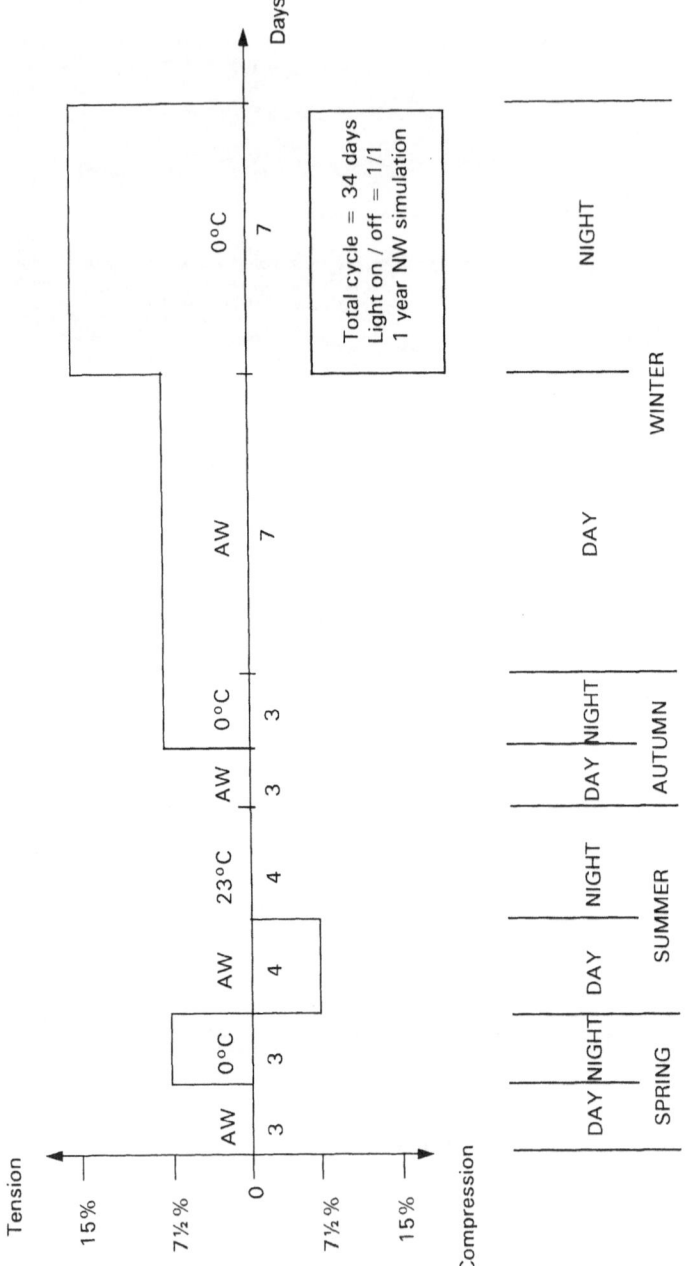

Fig. 8. London climate simulation.

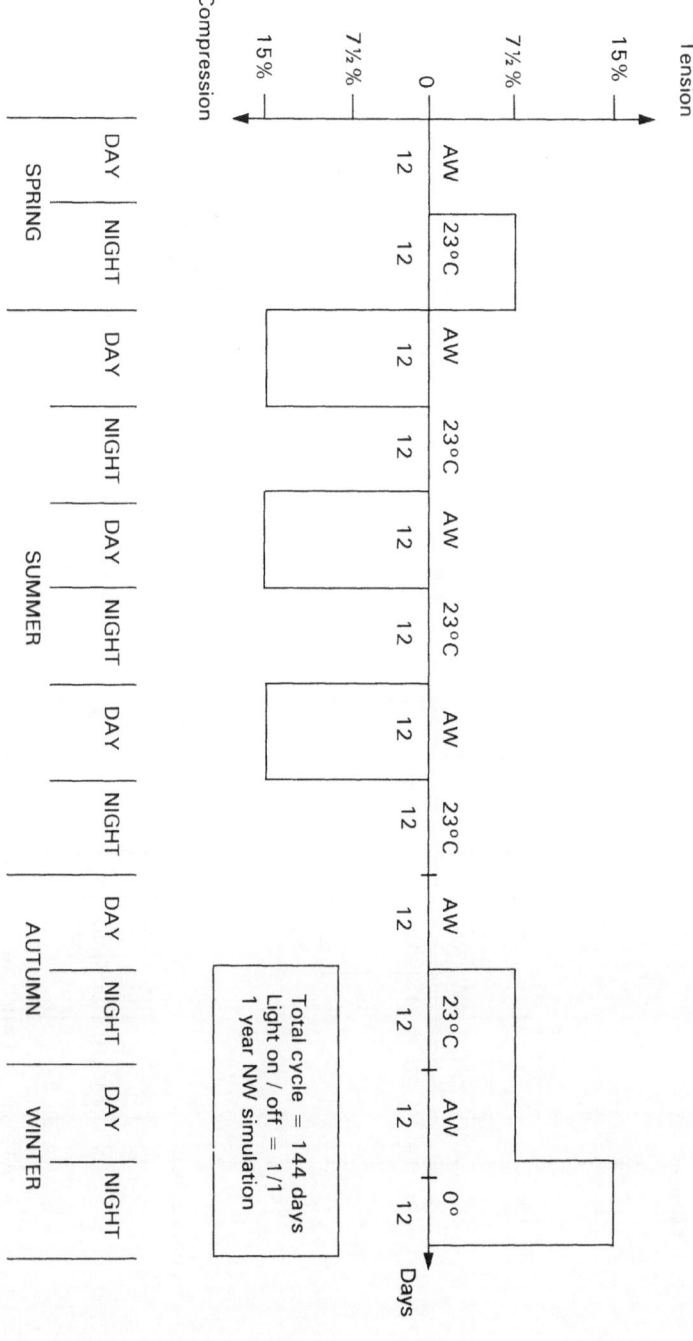

Fig. 9. Arizona (Phoenix), climate simulation.

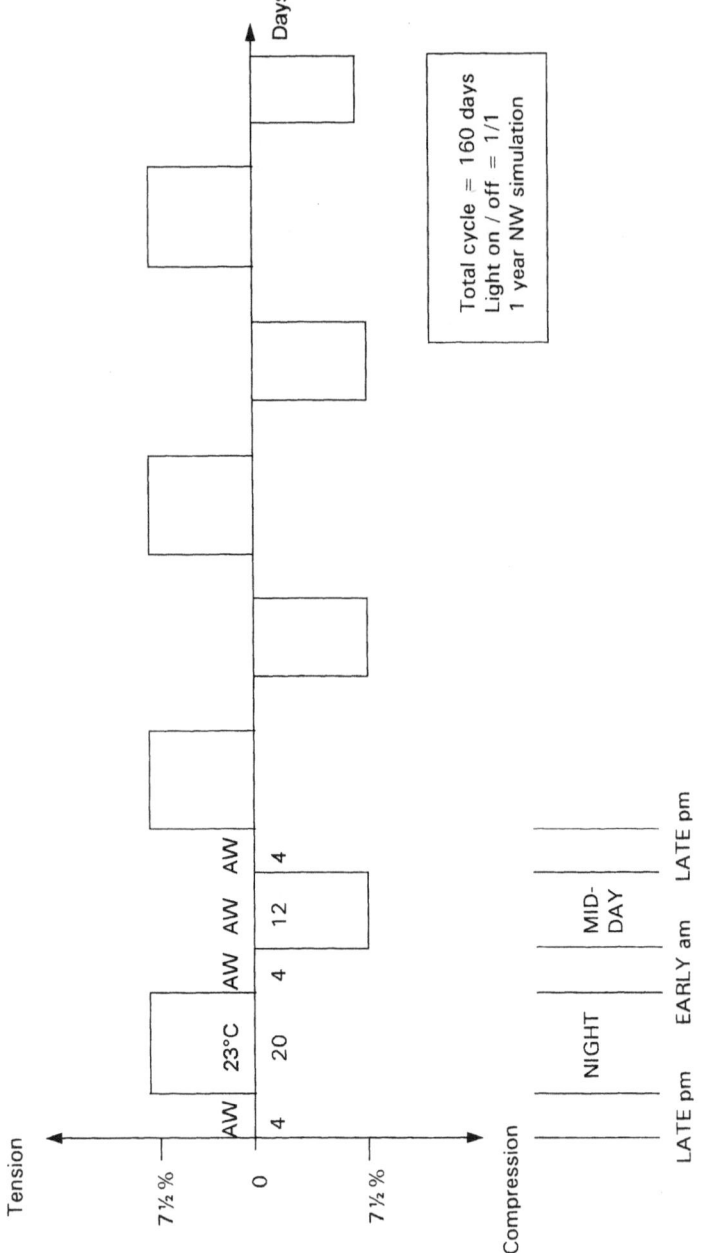

Fig. 10. Singapore climate simulation.

Table 2 Natural weathering/manual adjustments required for movement rigs

Period simulated	Month	Compression/ tension	% Of mean width	
			Silicone	PU/PS
	London			
	April/May	-	At mean width	
	June/July/Aug	Compression	33%	16%
	Sept/Oct	-	At mean width	
	Nov-March	Tension	16%	8%
	Phoenix			
	April	-	At mean width	
	May/June	Compression	33%	16%
	July/Aug	Compression	50%	25%
	Sept/Oct	Compression	33%	16%
	Nov/Dec	-	At mean width	
	Jan/Feb	Tension	16%	8%
	March	-	At mean width	
	Singapore			
Day (pm)	April/May/June	-	At mean width (exposed)	
Night	July/Aug/Sept	Tension	16% (covered)	8%
Day (am)	Oct/Nov/Dec	-	At mean width (exposed)	
Day (noon)	Jan/Feb/Mar	Compression	16% (exposed)	8%

mirror the seasonal variations shown by detailed climatic data. As it is difficult to impose rapid cycles over long test periods using manually operated test rigs, diurnal variations were initially omitted but then re-introduced in a simplified manner by the division of each seasonal phase. In the case of Singapore, there is very little seasonal variation in temperature. Consequently, this movement regime represents only diurnal changes.

The overall length of each cycle was set by correlating the total ultra-violet energy imposed during one complete cycle with that which would be received annually, on average, at each geographic location (a number of assumptions and approximations were necessary here).

Several general points should be made in connection with these movement cycles:

- As in the case of the accelerated weathering parameters, there was a wish to impose cyclic movement in a manner which acknowledged reality.
- Inevitably, a high degree of simplification was necessary - perhaps most notably in the representation of diurnal changes and the use of step-wise cycles.
- Such simplifications were accepted as there was also a need to determine whether varying test conditions induced detectable changes at significantly different rates,

giving a relationship between service-life and service conditions.

- The refinement of such regimes could possibly proceed in two directions: simplification based on existing knowledge of the relationship between sealant properties, joint type and likely performance; increased complexity and closer simulation of reality by the use of electrically powered test rigs - a difficult but feasible option.

5 Analytical procedures

The main intention here has been to convey the overall objectives of the project and to provide information on the exposure conditions utilised. Consequently, the analytical techniques employed will not be discussed but a summary is given in Table 3.

Table 3   Analytical techniques

Physical techniques	Chemical techniques
Tensile testing	XPS (Surface assessment)
- Bonded specimens	UV/visible spectroscopy
- Free films	FTIR and ATR-FTIR spectroscopy
Stress relaxation and	Micro-FTIR (transmission and reflection modes)
elastic recovery	Photo-acoustic-FTIR
Micro-swelling	NMR and ESR spectroscopy
DSC	Colorimetric measurements
Thermal mechanical analysis	Micro-fluorescence spectroscopy
Dynamic mechanical thermal	Labelling of degradation products by specific
analysis (DMTA/DLTMA)	chemical reactions
Gel permeation chromatography	
Surface energy (contact angles)	
(Finite element analysis)	

6 References

1. Woolman, R. and Hutchinson, A. (1994) *Resealing of Buildings - A Guide to Good Practice*, Butterworth-Heinemann.
2. Scott, J.L. (1983) *Programmed Environmental Testing - A PET Theory*, J. of Oil and Colour Chemists Association, Vol. 66 No.5, pp.129-132.
3. Karpati, K.K., Solvason, K.R. and Sereda, P.J. (1977) *Weathering Rack for Sealants*, J. of Coatings Technology, Vol. 49 No.626, pp.44-47.
4. Beech, J.C. and Turner, C.H.C. (1983) *Cyclic Joint Movements*, Building Research and Practice, Vol. 11 No.5, pp.287-291.

6 AGEING RESISTANCE OF BUILDING AND CONSTRUCTION SEALANTS (PART I)

A.T. WOLF
Sealants Science and Technology,
Dow Corning Corporation, Midland, USA

Abstract

The paper, which is the first in a series of papers on the ageing resistance of sealants, describes the effects of individual ageing factors on sealants. Ageing factors discussed in the paper are photochemical degradation and thermal loading due to sunlight; ambient heat and cold and the cyclic mechanical strain resulting from temperature fluctuations; water and water-vapour; oxygen and ozone; aggressive atmospheric pollutants; as well as micro- and macrobiological attack.

Keywords: Durability, ageing resistance, degradation, sealants

1 Introduction

People age, so do sealants. Ageing can affect the sealant's appearance, mechanical properties, or adhesion. As the mechanical and adhesive properties degrade with time, the sealant becomes less and less capable of withstanding the service conditions, until finally it fails. This failure, regardless of whether adhesive or cohesive in nature, causes the seal to leak and sets an end to the sealant's service life.

The capability of a sealant to resist ageing depends primarily on its generic type and to some extent on its formulation. It is not surprising, therefore, that ageing resistance is a property which varies widely between different types of sealants. Oleoresinous sealants may have service lives as low as three or four years; silicone sealants, on the other hand, have been shown to exceed 30 years of service life in weatherseal installations.

The ageing resistance or "durability" of a sealant is its most important property, since it is what ultimately determines the service life of the seal. Unfortunately, most sealant suppliers do not provide information on the long-term ageing characteristics of

Durability of Building Sealants. Edited by J.C. Beech and A.T. Wolf. © RILEM.
Published by E & FN Spon, 2–6 Boundary Row, London SE1 8HN, UK. ISBN 0 419 21070 9.

their products. A specifier, faced with the task of selecting a sealant for a specific application, should have some understanding of the changes that are likely to occur in these products as a result of ageing. Specifying a "low modulus polyurethane sealant or equivalent" is just not good enough. The specifier must have answers to the following questions: How much will the 25% modulus of the sealant increase within one, two or five years of outdoor weathering? How rapidly will the glass adhesion degrade when the sealant/substrate interface is exposed to sunlight? How rapidly will the sealant's adhesion to concrete degrade when permanently exposed to water? Specifiers should insist that sealant suppliers provide such information before deciding in favour of a specific product.

A dilemma, both sealant specifiers and suppliers are facing, is the fact that there is no universally accepted test standard for durability. Some movement capability test standards, such as ISO 9047 [1] or ASTM C-719 [2], have a certain amount of ageing built into their conditioning procedures. However, this ageing is only intended to complete the cure of the sealants; it is not rigorous enough to discriminate sealants that tend to fail after several years of service life. Some accelerated weathering tests, such as ASTM C-793 [3], only consider changes occurring on the sealant's surface. Such test standards, too, do not allow predicting a sealant's in service performance. Mechanical cycling of the joint specimens has to be part of any accelerated or natural weathering test in order to provide meaningful results. Tests which do not contain a mechanical cycling procedure are capable of reproducing surface changes that occur as a result of ageing, but the test results do not correlate well with the sealant's actual in-service performance.

The mechanical cycling of the test specimens ideally is carried out simultaneously to the natural or accelerated weathering. While this is easily achieved on outdoor weathering racks, it generally poses a problem in accelerated ageing method, since the test chamber of most weatherometers do not provide sufficient space for installing a cyclic stress/strain tester. If the mechanical cycling cannot be carried out simultaneously with the weathering, the second best alternative is to cycle the specimens repeatedly after ageing intervals.

Although over the last three decades many scientists have used accelerated weathering methods in studying the ageing performance of building materials, the selection of the ageing conditions remains more an art than a science. Mild conditions do not provide a sufficient acceleration effect, while too harsh conditions can produce artefacts not occurring under actual service conditions.

This paper tries to convey a basic understanding of the various environmental factors that can cause ageing of sealants. It does not attempt to convey the detailed knowledge required in interpreting ageing test results. The complexity of the information is more an indication of how little understanding has been gained of the phenomena termed "sealant ageing" than a result of the accumulation of knowledge from past studies. However, the author hopes that the RILEM TC139-DBS and ISO TC59/SC8 committees will succeed in developing an international test standard for the durability of elastic sealants, which, once published, will provide more comparable test results.

2 Environmental effects on building and construction sealants

Sealants in outdoor climates are exposed to the following environmental factors:

- sunlight (UV radiation, heat radiation)
- ambient heat and cold
- cyclic mechanical strain (compression/extension)
- water, water-vapour
- oxygen and ozone
- aggressive atmospheric pollutants (e.g. acid rain)
- micro- and macro-biological influences (e.g. fungi, termites)

In natural weathering, a number of environmental ageing factors can occur at the same time. For example sunlight, oxygen, ozone, heat and compression may exert their influences on the sealant simultaneously. The effect of these combinations does not simply equal the sum of the individual effects; it may be considerably greater or lower. In general, the different environmental factors reinforce each other, a phenomenon referred to as synergism. Experience has shown that sealants which resist the individual influences of ultraviolet light and moisture, often fail when exposed to both environmental factors simultaneously. An increase in temperature accelerates the rate of most chemical reactions. While photochemical reactions in general are not temperature sensitive, the subsequent chain reactions are temperature dependent. The combination of heat, water and ultraviolet light thus provides a strong synergism which greatly accelerates the ageing process. Some combinations of environmental effects can actually retard the ageing process. One example of a combination in which the resultant effect is retardation of ageing is the simultaneous influence of sunlight and airborne dust. The dirt deposit on the surface protects the sealant from further deterioration by ultraviolet light. This effect accounts for the lower ageing of sealants in industrial areas compared to rural areas with clean air.

The following chapters describe the ageing effects caused by the individual environmental factors. Part II in this series of articles will then deal with the synergetic effects between ageing factors and attempts at correlating accelerated and natural ageing.

2.1 Sunlight

Sunlight is solar electromagnetic radiation that reaches the earth's surface. Only a rather small band of wavelengths is capable of passing through the earth's atmosphere; upon impinging on the surface of the earth, sunlight consists of light visible to the human eye, infrared heat radiation and short wave, ultraviolet light. Since the energy of radiation is inversely proportional to its wavelength, ultraviolet light carries a higher energy than visible or infrared light. Approximately 42% of the solar radiation is in the infrared range (wavelengths >800 nm). This portion contributes to the heating of objects irradiated. Photochemical reactions, which cause degradation of polymeric binders, are not initiated by this type of radiation. Visible radiation with wavelengths between 400 and 800 nm makes up approximately 52% of the total solar radiation impinging on the earth's surface. Visible solar radiation contributes to heating of objects, but also is capable of initiating photochemical processes. Although ultraviolet light of wavelengths between 280 and 400 nm contributes only about 6% to the total solar radiation received by the earth's surface, it is responsible for many of the changes which occur, when polymeric materials are exposed outdoors. Ultraviolet light is capable of breaking chemical bonds, thereby inducing photochemical processes which have a decisive influence on the degradation of polymers. The contribution of ultraviolet radiation to heating of the irradiated object is correspondingly low. Table 1 shows the spectral distribution of radiation energy received on the earth's surface [4].

Table 1. Irradiance by Global Radiation (Perpendicular Incidence) (Data from [4])

Range	Wavelength (nm)	Irradiance (W/m^2)		Percentage of Total Radiation (%)	
0	<280		0		0
1	280-320	5		0.5	
	320-360	27	68	2.4	6.1
	360-400	36		3.2	
	400-440	56		5.0	
	440-480	73		6.5	
	480-520	71		6.3	
	520-560	65		5.8	
2	560-600	60	580	5.4	51.8
	600-640	61		5.5	
	640-680	55		4.9	
	680-720	52		4.6	
	720-760	46		4.1	
	760-800	41		3.7	
	800-1000	156		13.9	
3	1000-1200	108	329	9.7	29.4
	1200-1400	65		5.8	
	1400-1600	44		3.9	
	1600-1800	29		2.6	
4	1800-2000	20	143	1.8	12.7
	2000-2500	35		3.1	
	2500-3000	15		1.3	
5	>3000		0		0
Total:			1,120		100

Since the early 1970s, concerns were raised over increased ultraviolet radiation reaching the earth's surface due to stratospheric ozone depletion. More recently, both ground and satellite based data indicate a statistically significant long-term decrease in global stratospheric ozone levels of about 2.7±1.4% per decade. The rate since 1978 seems to have increased to about 4% per decade above North America and Europe [5,6]. There is yet little evidence of an accompanying wide-spread increase in terrestrial UV levels. Pickett [7] estimates, based on a mathematical model, that for total losses of less than about 10%, the loss of ozone (in percentage points) correlates linear with an increase (in percentage points) in terrestrial UV-B radiation levels. The ozone depletion will manifest itself in increased ultraviolet light intensity with a broad band centred around 310 nm with only very little absolute intensity increase for wavelengths shorter than 295 nm. Pickett also predicts that, unless sustained total ozone losses exceed 15% during the summer months, the effect on the ageing of most polymeric materials should be inconsequential, since a 10- 15% increase in UV dosage would be difficult to separate from the natural year-to-year variations due to changes in the weather pattern.

2.1.1 Photochemical degradation

Over time, exposure to the visible and ultraviolet component of sunlight can induce changes in a sealant's surface, bulk properties or adhesion. In pigmented sealants, photochemical ageing processes usually only affect the outer layer down to a depth of 0.2-1.5 mm, since most pigments either reflect or absorb visible and ultraviolet light. Sealants which are translucent to visible light may allow penetration of ultraviolet light up to several millimetres in depth, depending on the absorbance of the polymeric binder, additives or fillers.

According to Ashton [8], the short-wave components of sunlight can trigger the following photochemical reactions in organic construction materials:

- chain scission,
- crosslinking,
- formation of low-molecular compounds,
- modification of existing or formation of new functional groups.

Chain scission and crosslinking are by far the most important reactions to occur in sealants. As a rule both processes occur simultaneously in response to light, but one of them usually predominates for a given polymeric binder. If this is chain scission, the sealant softens over time. If crosslinking predominates, the sealant becomes brittle.

The *extent of ageing* induced by ultraviolet light is determined not only by the nature of the polymeric binder but also by the formulation of the sealant. For example, the rather poor inherent ageing resistance of polysulfide and polyurethane polymers can be improved by adding UV-stabilisers to the sealant formulation. However, the *type of ageing* observed is characteristic of the sealant's polymeric binder.

2.1.1.1 Effects of sunlight/ultraviolet light on the sealant surface

Building applications on our planet by definition imply that a sealant's surface is exposed to the earth's atmosphere. The ageing effects of sunlight on the sealant's surface are therefore synergetically confounded with the effects of atmospheric gases, especially that of oxygen. In practice the following changes may occur to a sealant's surface as a consequence of exposure to sunlight/ultraviolet light and the atmosphere [9]:

- changes in colour,
- wrinkles and crack formation perpendicular to the joint movement (elephant skin, orange peel, crazing),
- irregular Craquele crack formation (mud-cracking),
- chalking.

2.1.1.1.1 Colour changes

Changes in the surface colour of a sealant occur when either the pigments contained in the sealant formulation or the sealant itself are not sufficiently resistant to ultraviolet radiation. In the first case, the sealant's colour changes without any other signs of surface degradation being evident. In the second case, roughing or chalking of the sealant surface leads to a loss of colour intensity; an originally dark brown sealant can become light brown in the course of time - a phenomenon which has been observed fairly frequently with polyurethane and polysulfide sealants (see *chalking*).

It should be noted that some pigments, such as certain forms of titanium dioxide or iron oxides, can act as photochemical sensitiser in certain sealant formulations. For example, a polysulfide sealant formulation pigmented in black or grey may show good resistance to ultraviolet light, but may degrade rather fast, when pigmented in brown. The sealant manufacturer therefore has to evaluate all colour versions of a sealant product for their resistance to ultraviolet light.

2.1.1.1.2 Wrinkles and crack formation (orange peel, elephant skin)

Ultraviolet light may considerably accelerate the oxidation of the sealant's outer layer by atmospheric oxygen, causing a brittle and inelastic skin to form. According to Friberg [10], the increased formation of oxidised compounds in the surface of the sealant inhibits further oxidation, therefore the layer of brittle skin is usually only 0.5-1mm thick. Once the sealant is exposed to movement, the brittle surface cracks perpendicular to the movement and forms wrinkles. This phenomenon, typical of polysulfide sealants, is shown in Figure 1. The aged sealant shows a tough, wrinkled skin which is often referred to as "elephant's skin" or "orange peel" [11].

Fig. 1 Polysulfide sealant surface after ageing in Xenon-Tester for 5000 hours.

2.1.1.1.3 Craquele crack formation (mud-cracking), Chalking

A different type of surface ageing results when the sealant's polymeric binder is degraded into low molecular weight compounds. As a consequence of further weathering, these compounds are extracted from the outer sealant layer, leaving behind only inorganic fillers and some higher molecular weight degradation products. As a result, the sealant surface is starved of polymeric binder and shows little cohesion. Rubbing such a sealant surface with the thumb removes a whitish powder; this type of ageing therefore is termed "chalking". The extraction of the low molecular weight compounds from the sealant surface also results in a localised loss of volume. The shrinkage causes the surface to crack in an irregular Craquele pattern; a phenomenon also termed "mud-cracking". Mud-cracking, as shown in Figure 2, is typical of polyurethane sealants [11].

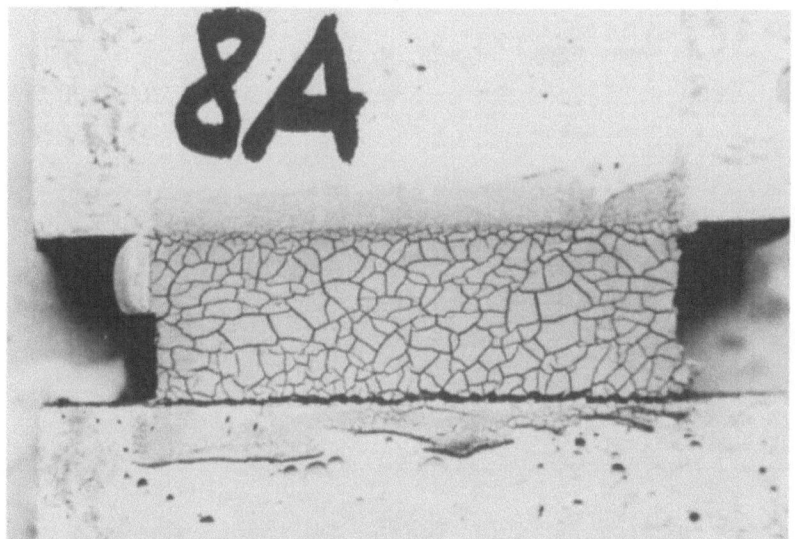

Fig. 2 Polyurethane sealant surface after ageing in Xenon-Tester for 5000 hours.

Some of the early polyurethane sealant formulations introduced in the 1970s showed such a high degree of mud-cracking and chalking that these sealants were eroded by the wind. Although today's polyurethane sealants have been much improved with the help of suitable ultraviolet light stabilisers, wind erosion is still a real concern in desert environments. Blazing sunlight and elevated temperature accelerate the oxidation of the sealant surface and the loss of low molecular weight compounds. Strong winds, carrying fine sand from the desert, remove the powdery sealant surface and expose deeper layers of the sealant to further degradation and erosion. Due to this "sand-blasting" process, desert winds may completely erode polyurethane sealants up to a depth of 20 mm within one or two years.

As mentioned before, the ageing phenomenon observed upon exposure to ultraviolet light is mainly determined by the nature of the polymeric binder, while the extent of ageing depends both on the nature of the polymeric binder and the formulation of the sealant. Within a class of sealants, resistance to ultraviolet light may vary one order of magnitude, depending on the sealant formulation. Given this limitation, the following assessment of the *resistance of sealants to ultraviolet light* is generally applicable: Ultraviolet light accelerates the resinification process of oleo-resineous binders used in glazing putties and low performance mastics. This resinification process causes a skin to form on the putty or mastic. If the skin is not protected with a coat of paint within a few weeks, the resinification process proceeds into the depth of the sealing compound, causing it to severely harden. Butyl mastics or modified putties show moderate resistance to ultraviolet light. The appearance of acrylic sealants is little influenced by exposure to ultraviolet light (Figure 3), although, depending on the sealant formulation, yellowing may occur in some cases [9,11]. Exposed to sunlight, acrylic sealants tend to harden, which may be attributed as much to thermal ageing as to degradation due to ultraviolet light.

Fig. 3 Polyacrylate sealant surface after ageing in Xenon-Tester for 5000 hours.

The tendency of polyurethane sealants to develop mudcracks has been reduced during the 1980s with the help of stabilisers, but this ageing phenomenon may still be observed to a varying degree for most one-part polyurethane formulations [9,11]. Polysulfide sealants develop a tough, wrinkled skin when exposed to ultraviolet light [9,11]. Silicone sealants and gaskets display excellent resistance to ultraviolet light [9,11]. Even after many years of heavy exposure to sunlight, the surface of silicone sealants does not display any significant changes (Figure 4).

A recent model study into the photo-oxidative stability of polydimethylsiloxanes indicates that the excellent stability of these materials may be somewhat reduced by the introduction of dimethylene groups (C-C bonds) into the polymeric network, as is the case with addition cured siloxane materials [12]. Because of its higher cost and propensity to catalyst deactivation, the addition cure system is generally not utilised for silicone sealants. However, the polymeric backbone of so-called "modified silicone sealants" (silyl modified polyethers), which is prepared via hydrosilylation of alkylene endblocked polyethers does contain dialkylenesilicon groups (C-C-Si bonds), which are susceptible to photo-oxidation. Since the polymeric backbone of these sealants is inherently instable towards UV exposure, any instability resulting from the introduction of dialkylenesilicon groups is confounded by that of the polymeric backbone.

Fig. 4 Silicone sealant surface after ageing in Xenon-Tester for 5000 hours.

Photo-oxidation of polychloropren (CR) rubber occurs rather slowly. However, since it results in the formation of additional crosslinks, CR gaskets harden upon exposure to ultraviolet light [13]. EPDM polymer does not carry any double bonds. Gaskets formulated on this polymer base therefore displays better resistance to ultraviolet light than CR rubber [13]. When submitted to mechanical strain, small fissures form in the brittle surface of aged CR and EPDM rubber gaskets, a phenomenon termed "crazing". In order to improve their ultraviolet light resistance, EPDM and CR rubbers are generally pigmented in black, since carbon black is a very efficient ultraviolet light absorber.

2.1.1.2 Effects of sunlight/ultraviolet light on the sealant's glass adhesion
The above discussion of changes induced by ultraviolet light involved only those occurring on the surface of the sealant. However, in the case of cured-in-place glazing or insulating glass sealants, ultraviolet light may also affect the sealant's glass adhesion, since the sealant/glass interface is directly exposed to sunlight. Silicate glass is not translucent to ultraviolet light with wavelengths below 280 nm [14]. However, as Gjelsvik [15] already pointed out in 1975, photon energies of ultraviolet light with wavelengths above 280 nm are quite sufficient to split a number of chemical bonds such as C-N, C-Cl, C-O, C-C or C-H. Compared with the total number of chemical bonds formed during cure, adhesion of the sealant to the substrate stems from a relatively low number of bonds. It is therefore obvious that the irreversible destruction of a

comparable low number of chemical bonds in the boundary area between sealant and substrate is sufficient to evoke a permanent loss of the sealant's adhesion.

Most organic cured-in-place sealants exhibit poor inherent adhesion to glass which is rapidly lost upon exposure to ultraviolet light. In order to achieve good adhesion on glass, a silane needs to be added to the sealant formulation or applied as a primer prior to sealant application. The adhesion is then established by the silane providing the link between the glass surface and the organic sealant polymer. Since the silane itself is very resistant to ultraviolet light, it also somewhat improves the adhesion of the organic sealant in this respect. Despite the use of a silane primer, however, the organic sealant still fails after prolonged exposure to sunlight, leaving a thin layer of sealant on the glass. This failure mode indicates that the bond cleavage actually occurs in the organic sealant polymer and not in the silane primer. Contrary to that of organic sealants, the glass adhesion of silicone sealants remains virtually unaffected by prolonged exposure to ultraviolet light.

2. 1.2 Thermal loading

Certain wavelengths of solar radiation impinging upon a building facade are absorbed and converted into heat energy, while others are simply reflected. The surface temperature of a facade will, thus, depend on the amount of incident radiation energy and the absorbance of the facade surface. This law of physics has some simple consequences: First, the darker the facade surface, the more radiation energy that is converted into heat. Second, the more sunlight that hits the facade surface, the more heat that is generated. On the northern hemisphere, for instance, the South facing side of a building will get hotter than the North facing one. Building surfaces, which are inclined towards the sun, receive more solar radiation energy than vertical surfaces. The highest radiation energy is received when the sunlight impinges perpendicular on the building surface. One of the first studies into the surface temperature of building elements was carried out in 1968 on different coloured plasters applied to a west-facing, hollow pumice block wall [16,17]. The darker the colour of the pumice, the higher the measured maximum surface temperature was.

Changes in solar irradiance over the course of a day result in temperature variations of a building facade, causing the building joints to expand and contract, which imposes cyclic mechanical strain on the sealant. Thermal loading of building elements by solar radiation thus causes the same kind of stress on building joint sealants as ambient temperature variations do. However, it should be noted that a building facade frequently is not in thermal equilibrium with the ambient climate: Under solar irradiation, temperatures of dark coloured building constructions with low heat capacity, such as aluminium curtainwalls, exceed the ambient temperature by far. The effects of thermal loading on sealants will further be discussed below in chapter 2.2.2.

2.2 Cyclic mechanical strain, ambient heat and cold

2.2.1 Cyclic mechanical strain

As discussed previously, variations in ambient temperature cause building joints to move, imposing cyclic mechanical strain on the sealants. Such strain, which recurs daily and varies as a function of the season, may cause degradation of the sealant's appearance, adhesion and mechanical properties. Karpati [18-28] has intensively researched the behaviour of sealants under cyclic mechanical strain and regards it as the most important ageing factor. One of Karpati's studies [26] showed that mechanical cycling of a cured two-part polysulfide sealant at ambient laboratory temperature and in absence of any other ageing factor was capable of reproducing the same permanent deformations and consequent failure mechanisms as observed in outdoor weathering under forced mechanical cycling.

If a cured sealant is permanently subjected to cyclic movements, cracks may gradually form on its surface at a right angle to the direction of the stress. The number and amplitude of the imposed movements, the speed (frequency) with which they occur and the temperature in the sealant during the movement cycle are decisive factors in terms of the scope of mechanical ageing. These parameters are interrelated to the extent that an increase in the speed of deformation increases the stress in the sealant. If the sealant is subjected to movements greater than its movement capability, the cracks in the sealant surface grow and may finally culminate in a cohesive failure. Adhesive failure may also result when the strain at the sealant/sub- strate interphase exceeds the sealant's bond strength.

2.2.2 Heat

Heat can trigger both chemical as well as physical effects in sealants. Higher temperatures accelerate most chemical processes, so that ageing caused by other environmental factors proceeds more quickly at higher temperatures. According to Arrhenius' law, the rate of chemical reactions doubles with a temperature increase of 10°C.

Thermoplastic sealants such as solvent acrylics soften when heated. This is a result of the loss of weak bonds between the polymer chains. If the heating period is brief, most of the effects on thermoplastic sealants are reversible, assuming that the temperatures in question were not so high as to initiate polymer degradation [29].

Elastic sealants generally contain plasticisers which may evaporate at higher temperatures, causing the sealants to embrittle with time. The extent of the plasticiser evaporation obviously depends on the vapour pressure of the plasticiser at a given temperature, but also on its compatibility with the sealant's polymeric matrix. Most organic elastic sealants (polysulfides, polyurethanes, silyl modified polyethers, etc.) contain low molecular weight plasticisers, such as dioctylphthalate. These sealants show significant plasticiser evaporation at temperatures above 80°-100°C. Silicone sealants, if plasticised, generally contain polymeric trimethylsilyl endblocked siloxane plasticisers, which do not show a significant evaporation rate until temperatures above 150°C are reached. Recently, some manufacturers have commercialised silicone sealants containing non-silicone plasticisers, such as organic waxes, paraffines, or phosphate esters. These formulations generally display a lower heat resistance than regular silicone sealants.

For certain applications in the industrial or construction sector, where sealants are subjected to extremely high temperatures, such as solar collectors, particular attention must be paid to the evaporation of plasticisers or other constituents of the formulation. Not only may the sealants embrittle over time, the low molecular weight plasticiser may also condense at colder locations and thus detrimentally affect the efficiency of the solar collector [30].

Exposure of cured elastic sealants at moderately elevated temperatures (up to about 70-80°C) generally does not lead to any degradation of the polymeric backbone.

However, even moderate temperatures provide sufficient energy to the polymeric network to trigger formation of additional crosslinks, a phenomenon termed as "post-cure". The post cure manifests itself in an increase in moduli and surface hardness (durometer) and a decrease in maximum elongation. The tensile strength of the sealant typically remains almost constant, since the increase in the sealant's modulus is approximately compensated by the reduction in maximum elongation. Almost all sealants exhibit a post-cure phenomenon, however, the degree of post-cure experienced at a certain temperature depends on the sealant's polymer type and formulation as well as the duration of the exposure.

According to Burström [29], heat induced ageing exhibits the highest influence upon cured-in-place sealants. Sealants with a polysulfide or polyurethane base significantly post-cure upon heat ageing. For example, the 50% modulus of a one-part polyurethane sealant may increase up to 350% within 100 days exposure at 70°C [31]. However, as mentioned above, the extent of the post-cure is highly dependent upon the sealant's formulation. A study by Beech and Beasley [32] confirmed Burström's finding that heat ageing causes a post-cure in most sealant types, including silicone sealants. However, for certain polyurethane and polysulfide formulations, these authors found a reduction in 25% moduli after 52 weeks of 70°C storage versus the controls stored at room temperature. Further analysis of Beech's and Beasley's findings shows an initial increase in the moduli of these sealants over the first 8-16 weeks of heat exposure. Apparently, storage at 70°C caused these sealants to initially post-cure but also provided sufficient energy for polymer degradation to occur over extended periods of time.

As mentioned before, temperatures of 40-70°C are sufficient to induce post-cure in sealants. The higher the temperature, the faster the sealant's mechanical properties reach a plateau. Temperatures above 70°C are generally not considered in studying post-cure, since polymer degradation reactions may occur in organic sealants at temperatures above 80-120°C. The temperature threshold, at which noticeable polymer degradation reactions occur, is determined primarily by the sealant's polymer type, secondarily by its cure chemistry, and tertiary by its formulation, especially by filler type and level. Since polymer degradation and crosslinking reactions compete at elevated temperatures, either significant softening or hardening may result.

How can one determine the upper service temperature limit for a sealant? For most applications it is useful to know whether and to what extent changes occur that affect modulus, tensile strength and elongation at break, as well as adhesion to specific substrates. These changes can be determined by preparing standard ISO 8339 [33] $12 \times 12 \times 50$ mm^3 test samples and storing them at various elevated temperatures. After heat storage, the test specimens are pulled to break in a tensile tester. In order to identify trends in the change of material properties, the tensile test has to be carried out after various heat storage intervals; this requires the preparation of a sufficient number of test specimens.

Table 2 provides ranges of the upper service temperature limits for various sealant types based on exposure to hot, dry air [13,34]. This upper service temperature assessment is based on a guideline developed earlier by this author that storing a cured sealant at these temperatures in an oven with forced ventilation over a period of 6 months should not cause the modulus, elongation at break or tensile strength to change by more than 50% of their initial values. The widths of the temperature ranges account for differences in product formulation. Silicone sealants display the highest resistance to thermal degradation since the energies of the Si-O and Si-C bonds are considerably higher than that of a C-C bond, which is present in organic sealants. Service temperatures of up to 180-200°C can be achieved with acetoxy, aminoxy, or oxime cure silicone sealants, while alkoxy or benzamide cure silicone sealants do not withstand service temperature above 120-150°C for prolonged periods of time [34]. Use of special fillers, such as iron oxide, improves the heat resistance of silicone sealants even further, and allows formulation of products with upper service temperature limits up to 220°C. Yang [35] proposed that these fillers behave as secondary antioxidants in silicone elastomers by scavenging the free radicals formed. If the exposure time is limited to a few days or hours, these special silicone sealant formulations will resist temperatures up to 250°C or 300°C, respectively. In general, fillers which have been chemically pacified with a surface treatment provide better heat stability in sealants than untreated fillers.

Table 2. Ranges of upper service temperature limits for various sealant materials

Polymer Base	Cure System	Upper Service Temperature (°C)
EPDM		120
Chloroprene (CR)		90
Polysulfide		70 to 90
Polyurethane		80 to 90
Acrylics	Solvent	50 to 70
	Latex	80 to 120
Silicone	Acetoxy	150 to 220
	Alkoxy	100 to 150
	Aminoxy	150 to 200
	Benzamide	80 to 120
	Oxime	120 to 200
	Latex	120 to 200

Briefly storing a freshly cured sealant at somewhat elevated temperatures, 2-3 days up to several weeks at 70°C for example, generally accelerates the build-up of full adhesion to the substrate. This effect was described by Karpati in connection with both heat and outdoor storage [23]. Extended storage of the sealant/substrate sample at temperatures above the upper service limit always results in a degradation of adhesion.

Thermal degradation will not only affect the mechanical properties and adhesion of sealants, but also cause decomposition gases to form. The formation of decomposition gases is often ignored in determining the upper service temperature limit of sealants, but can become a major problem on a building site. The University of Kassel, Germany, can be cited as an example of such an occurrence, where students started to complain about nausea after the university's high temperature heating system had been sealed with a two-part polysulfide sealant in the autumn of 1979.

2.2.3 Cold

All sealants, regardless of their polymer base and formulation, gradually stiffen when the ambient temperature is lowered. The extent of this stiffening is primarily determined by the nature of their polymeric binder. If the temperature is lowered further, the sealant suddenly loses its elasticity and becomes brittle. The temperature, at which this sudden embrittlement occurs, is related to the glass transition temperature (T_g) of the polymeric binder. The higher the T_g of the polymeric binder, the earlier the sealant will stiffen as the temperature is lowered. The influence of cold on sealants is reversible; the original flexibility returns at higher temperature again. Nevertheless, an adhesive or cohesive failure of the seal may result from the stiffening of the sealant as the joint expands at low temperatures.

It is important to note that the glass transition temperature of a sealant is not identical with its lower service temperature limit. In most applications, the sealant is expected to accommodate movements at low temperatures. According to this author, the lowest temperature, at which the sealant is still capable of accommodating the desired movement, should be considered the lower service temperature. As with the upper service temperature limit, no standardised test method exists to determine the lower service temperature limit. However, existing test standards can be modified intelligently to allow an educated guess whether a sealant will perform at low temperatures or not. In order to do so, one needs to consider the service conditions, the sealant will encounter. How low will the temperature drop during winter and how long will it prevail? How much movement is the sealant expected to accommodate at the low temperature? How fast will the movements be? The specifier needs to provide some estimates for these parameters, in order to enable the sealant manufacturer to conduct meaningful tests.

For most building applications, the sealant temperature is not likely to fall below -20°C or to exceed +70°C. This is also the temperature range considered in movement capability tests for cured-in-place sealants. A sealant passing ISO 9047 [1] can be assumed to have a service temperature range of -20°C to +70°C. If the sealant fails the ISO 9047 test because of excessive stiffening at -20°C, the ISO 9047 test could be repeated with a less demanding low temperature limit. The temperature which allows the sealant to pass the ISO 9047 test should then be considered its lower service temperature limit. If a sealant is likely to be exposed to very low service temperatures, as occur in refrigerated warehouses or in Arctic climates, the specifier should ask the sealant supplier to provide ISO 9047 test results based on the lowest service temperature expected.

Table 3 shows lower service temperature ranges for various sealant types based on this author's experience with sealant performance in various flexibility and movement capability tests. Because of their excellent low temperature flexibility, silicone sealants are specified for extreme low temperature applications, such as remote housing complexes at Arctic oil drilling sites. Polyurethane sealants generally also display good low temperature performance, and are therefore specified in applications involving temperatures as low as -40°C, although significant hardening already occurs at temperatures below -20°C. A high bond strength and, thus, detailed preparation of the substrate surfaces (generally with primers) is required in order to achieve acceptable joint movement capability with polyurethane sealants at low temperatures. While polysulfide sealants retain their flexibility down to comparably low temperatures, their moduli increase even

more than those of polyurethane sealants. It has also been reported that some polysulfide sealants tend to develop crystalline regions, if maintained at low temperatures for
several days [36]. This crystallisation does not occur spontaneously. The duration of
the low temperature may therefore have an influence on the flexibility and movement
capability of these sealants at low temperature. Sealants based on acrylate polymers
generally show poor low temperature flexibility. This is especially true for the solvent
based acrylics and the low price/low performance acrylic latex sealants. Special test
standards, such as ASTM C-734 [37], were developed to assess the cold temperature
flexibility of acrylic latex sealants after artificial ageing.

Table 3. Ranges of lower service temperature limits for various sealant materials

Polymer Base	Cure System	Lower Service Temperature (°C)
EPDM		-20
Chloroprene (CR)		-20
Polysulfide		-20 to -30
Polyurethane		-30 to -40
Acrylics	Solvent	0 to +10
	Latex, pigmented	-5 to -20
	Latex, clear	+15 to -15
Silicone	Acetoxy	-40 to -60
	Alkoxy	-40 to -60
	Aminoxy	-40 to -60
	Benzamide	-40 to -60
	Oxime	-40 to -60
	Latex	-30 to -40

2.3 Water and water-vapour

Water is omnipresent in our environment, whether in the form of airborne humidity,
rain or dew. Sealants used in building or construction joints may be exposed to a wide
range of service conditions in which water may have a significant effect upon their performance. Service conditions range from insulating glass edge seals, which are generally exposed to water vapour only; over joints in the external walls of buildings, where
exposure to water is usually intermittent and of short duration; to fully immersed joints
in water retaining structures, where the sealant must withstand prolonged exposure to
water. Water or water vapour can exert physical and chemical influences on sealants.
The first are due to the capability of water to swell and leach sealants, while the second
comprise any damage caused by hydrolytic reactions.

2.3.1 Physical effects: swelling and leaching

Immersing a cured sealant sample in water or exposing it to a high relative humidity
will cause water molecules to diffuse into the sealant. The water molecules are deposited in the free volume of the polymeric network and, as a result, macroscopic swelling

of the sealant occurs. The absorbed water also acts as a plasticiser, hence the sealant's indentation hardness, modulus and tensile strength decrease, while its elongation at break may or may not increase, depending on the nature of the polymeric binder and the sealant formulation [10].

The swelling of the sealant due to water absorption is reversible; a subsequent drying period will entail the loss of water. Drying of the sealant's surface layer leads to a volume contraction, which is hindered, however, since the underlying layers are still swollen. The drying process thus induces a tensile stress in the sealant surface which may cause subsequent cracking. Swelling and drying per se represent a substantial stress on the sealant. This stress, however, may be further aggravated by photochemical ageing processes which embrittle the sealant surface and thus increase the likelihood of cracking under tensile stress.

Liquid water can also leach formulation constituents such as plasticisers, pigments or fillers from the sealant, a process which results in a loss of weight upon drying. Leaching may also occur due to hydrolytic breakdown of the polymeric binder. Sealants which cure only physically by evaporation of the dispersing agent can be redispersed by water. Water- or solvent-borne acrylics therefore cannot be used in applications involving long-term water immersion.

The degree of water absorption and, consequently, swelling is highly dependent upon the experimental parameters, such as the pre-conditioning of the sealant, which affects its state of cure, the duration of the water immersion, and the temperature of the water bath. It is, therefore, important to establish a comparable state of cure in sealant samples for comparative studies. At low to moderate water temperatures, the water absorption, as the result of Fick's diffusion law, is initially proportional to the root of the time. After a protracted period, typically 4-12 months, most sealants approach a saturation value. The rate at which this saturation state is approached is controlled by the temperature dependency of water diffusion into the sealant, which follows Arrhenius' law. The maximum water absorption arising in saturation equilibrium is also temperature dependent.

The degree of water absorption and swelling also depends on sealant specific parameters, such as the polarity of the sealant's polymeric binder and plasticiser components, the hydrophilicity of the filler used, which is affected by the nature and degree of filler surface treatment, the type and level of the adhesion promoter, and the crosslinking density of the elastomeric network [9]. Organic cured-in-place sealants exhibit a high degree of swelling when exposed to liquid water, while silicone cured-in-place sealants swell very little; their water absorption is approximately 1/3 to 1/10 that of organic sealants tested under identical conditions [38-44]. In a recent study on various polysulfide sealant formulations exposed to 95% relative humidity in air at 60°C for 14 months, Lowe et al. [45] found that the formulations studied absorb large amounts of water with weight increases ranging from 45 to 92%. The tendency towards leaching is very dependent on the formulation of the sealant; typically polysulfide and acrylic sealants show a higher propensity towards leaching than do sealants with a polyurethane or silicone base.

2. 3.2 Chemical effects: completion of cure, hydrolysis of chemical bonds, and loss of adhesion

Since the cure of most one-part sealants is triggered by water diffusing into the sealant, exposure to water or water vapour can lead to the formation of additional crosslinks, if the sealant has not had the chance to completely cure prior to the exposure [44]. This effect is most pronounced with slower curing one-part sealants, such as polysulfide, polyurethane or silicon modified polyether sealants, since the cure of these one-part sealants is limited by the low diffusion rate of water vapour into the sealant and eventual cure by-products from the sealant. Because of the plasticising nature of water absorbed in the sealant, the additional cure induced by water often is not noticed until the sealant is allowed to completely dry at the end of the water exposure.

Water is capable of hydrolysing chemical bonds within the sealant's polymeric matrix and between the sealant and the substrate. Hydrolysis of chemical bonds within a sealant's polymeric matrix occurs even at room temperature. However, given a temporary or intermittent exposure to water, hydrolysis at lower temperatures is generally quite limited and does not manifest itself in significant changes in the sealant's macroscopic behaviour. Beyond a certain, polymer dependant, elevated temperature threshold, however, the hydrolytic breakdown is always irreversible. The degradation manifests itself by water being continually absorbed into the sealant's polymeric matrix without reaching an equilibrium saturation limit. For silicone and polysulfide sealants, no significant polymer degradation reactions were noticed up to 90°C [38,46].

A sealant's adhesion to a substrate stems from a relatively low number of chemical bonds. It is therefore not surprising that hydrolysis of these bonds, even at room temperature, can strongly affect a sealant's adhesion. The speed with which adhesion is lost depends on the number of such bonds and their binding energies. For most sealants this process is irreversible; their adhesion is permanently lost when exposed to water for a prolonged period of time. Other sealants are capable of recovering their adhesion upon drying, provided the adhesion had not been completely lost during the water exposure. Whether the adhesion suffers reversible or irreversible damage by exposure to water depends on the nature of the polymeric binder and the formulation of the sealant, especially the nature of the adhesion promoter used.

The loss of adhesion due to hydrolysis of chemical bonds can be accelerated by immersing the sealant into water at elevated temperature [47]. For silicone sealants it has been shown that the temperature dependency of the adhesive bond hydrolysis closely follows Arrhenius' law [48,49]. The author expects other elastic sealant types to behave basically in the same manner, since the adhesion of organic sealants is achieved via siloxane bonds formed by organosilane adhesion promoters. Since the activation energy of the hydrolysis reaction depends on the nature of the substrate and to some extent on the type of adhesion promoter used in the sealant formulation, acceleration factors correlating room temperature with elevated temperature water immersion can only be derived for a given sealant/substrate combination. For a specific structural glazing sealant applied to float glass, Iker at al. [50] found that one week water immersion at 55°C closely reproduces adhesion loss observed after eight week water immersion at 23°C.

In a recent paper, Donaldson [51] showed that the service life of a silicone elastomer immersed in hot water depends on the isoelectric point of the substrate surface under the specific immersion conditions. The isoelectric point (IEPS) of a substrate surface under water is the pH at which the surface has no electric charge. In Donaldson's study, lap shear test samples made from the silicone elastomer and various metallic and ceramic substrates were boiled in buffer solutions at pH 4, 7, and 10 until they could no longer withstand a certain force applied for 10 seconds. The service life was defined as the time to failure for each elastomer/substrate combination. When plotting the logarithm of the time to failure (in days) versus the IEPS of the various metallic substrates studied, Donaldson found a linear relationship for pH 4 and 7, and a rather complex (parabolic) relationship for pH 10. Donaldson postulates that the service life of a given elastomer/substrate combination at a given immersion condition can be predicted based on the IEPS of the substrate surface.

Lowe et al. [45] studied the water resistance of adhesive bonds between glass and polysulfide sealants by exposing cured sealant/glass joints to 95% relative humidity and 60°C in air. All model formulations were based on glycidoxypropyltri- methoxysilane as adhesion promoter. The authors found the adhesion of all sealant formulations to degrade with time, however, the adhesion of formulations containing higher levels of (benzyl phthalate) plasticiser degraded significantly faster. Most of the adhesion strength was lost in the initial months of exposure and adhesive strength values levelled off after about 10 months. The authors also found that the rate at which the loss of adhesion occurred was faster than could be accounted for based on the rate at which water was diffusing into the sealant. Based on contact angle measurements, the authors estimated the thermodynamic work of adhesion of the polysulfide to glass bonds to be positive in dry conditions, but to be slightly negative in presence of water, which indicates that water may displace the sealant from glass. Interestingly, they also found that the plasticiser cannot displace the polysulfide sealants from glass, as the work of adhesion in its presence remained positive.

The hydrolysis of chemical bonds can be accelerated by alkalis which stem, for example, from the substrate. Hence a primer must be applied to highly alkaline substrates, such as fresh concrete, providing a barrier between the alkali and the sealant.

Two papers by Mansfield try to assess sealant lifetime in a water immersed environment considering both physical and chemical effects by studying changes in moduli and adhesion of tensile adhesion joints [44,52] .

2. 4 Oxygen and ozone

Oxygen causes gradual oxidation and embrittlement of organic sealants, especially in conjunction with higher temperatures and/or simultaneous exposure to UV radiation. Generally noticeable oxidation occurs only in the outer layer of the sealant, causing the formation of a brittle skin on its surface. For example, mud-cracks, formed upon outdoor ageing of one-component polyurethane sealants, are caused by the photochemical oxidation of the sealant surface.

Although the concentration of ozone in the earth's atmosphere is very low, the influence of ozone on the ageing behaviour of sealants cannot be ignored, since ozone exhibits a chemical reactivity about 200-500 times greater than atmospheric oxygen. Both oxygen and ozone therefore contribute to the oxidation of organic sealants.

The only systematic study of the effect of ozone on elastic and plastic sealants is that of Burström [29]. In this study, Burström found that, as for all elastomers, the damage evoked by ozone is most severe, when the sealants are simultaneously subject to a mechanical prestress. The damage then is proportional to the tension in the sealant and to the ozone concentration. Sealants in which tensions are more quickly relieved because of their plastic deformation behaviour consequently sustain less damage.

2. 5 Aggressive atmospheric pollutants

Aggressive atmospheric pollutants include chlorine, hydrogen chloride, sulphur dioxide as well as nitrogen oxides, the latter two pollutants being the main constituents of acid rain. Depending on the concentration of these gases, they can subject sealants to considerable chemical stress. A recent paper by Keshavara and Tock [53] studied the effect of simulated acid rain (pH 3) on a structural silicone sealant. Initially, an increase in the sealant's modulus was observed, which was attributed to the additional crosslinking triggered by the presence of moisture. Continued exposure, however, resulted in a decrease in the sealant's elastomeric properties.

Acid rain shows only moderate acidity (pH 3.5 to 5.0). However, some of the dissolved acids are of lower volatility so that evaporation of water from a rain droplet sitting on a horizontal surface causes it to become more corrosive. This concentrated acid (pH 1.5 to 2.0) can chemically react with and dissolve calcium carbonate fillers in sealants, rendering their surface more prone to mould fungus growth. However, no detailed studies of the effect of acid rains on sealants are available to date.

2. 6 Micro- and macrobiological influences

Building sealants can be attacked by microbiological agents, such as bacteria and fungi, and macrobiological agents, such as insects, rodents, and birds. Microbiological attack is to a large extent chemical in nature, while macrobiological attack is primarily physical. The susceptibility of building sealants to biological attack not only depends on the nature of the polymeric binder but also on the sealant's formulation, especially the type and level of plasticiser used. Plasticisers may impart to an otherwise inert polymer system their own biological properties [54]. The extent of their influence is governed by their susceptibility to biological attack as well as their biological availability, which in turn is influenced by their compatibility with the cured sealant's polymeric matrix, their migration rate to the sealant's surface and their volatility. Not all formulation ingredients increase a sealant's susceptibility to biological attack. Some additives are biologically inert, others may even increase bioresistance, as, for example, organic tin catalysts do, which assist in fungal resistance [54].

2.6.1 Microbiological attack

Divergent conclusions are to be found in the literature on the microbiological resistance of sealants. The confusion is caused by the fact that in many publications the formulations of the sealants studied are either unknown or not disclosed so that the effect of the polymeric binder cannot be separated from that of the other formulation ingredients.

2.6.1.1 Mould fungus growth

Mould fungus growth constitutes the most frequent cause of microbiological damage to sealants [55-57]. While mould may not cause physical deterioration of the sealant, it detrimentally affects the service value of sealants from an aesthetic and hygienic perspective. The discoloration of the sealant surface is only a by-product of the mould growth, caused by the dyes formed by the mould's metabolism. Once the mould has grown on the sealant surface for some time, the dyes have penetrated so deeply into the sealant's surface that the stains no longer can be removed by conventional cleaning agents.

The importance of fungi as deteriorative agents is mainly the result of the production of enzymes that are capable of breaking down organic materials. Certain environmental conditions are essential for growth and degradative activity. These include an optimal temperature, the presence of nutrient materials, and high humidity.

Generally speaking, synthetic polymers are more resistant to fungi than natural binders. Polysulfide, polyether and acrylic polymers are reported to be inherently resistant against mould fungus degradation [58]. Because of its inorganic backbone, silicone polymer cannot be broken down by mould fungi and therefore cannot supply nutrients for mould growth [58]. Mould fungi are therefore unable to settle on sealants formulated from these polymers unless additional organic nutrients in form of plasticisers, organic pigments, filler treatments, or an organic surface contamination are available. Formulations containing fatty acid residues are especially susceptible to mould fungus growth [59]. The fatty acid residue may stem from drying oil binders used in oleoresineous putties (linseed or tung oil), from non-slump additives used in polyurethane sealant formulations (castor or dehydrated castor oil), or from filler treatment (stearate) in calcium carbonate filled sealant formulations. As mentioned before, some plasticisers are readily utilised by fungi as a source of nutrients [59]. Destruction of the plasticiser results in surface tack, embrittlement, and discoloration of the sealant. Ester plasticisers are generally more prone to attack by fungi than alkyl phthalates or phosphates [59,60]. Cure by-products of sealants may either act as a nutrient for fungi, such as the fatty acid by-product of the octoate silicone cure system, or act as a biocide, such as the hexylamine by-product of the amine silicone cure system. Excluding the temporary effect of cure by-products on mould growth, it is highly likely that organic plasticisers and/or organic surface contaminations are the primary source of nutrients for mould fungi, regardless of the polymeric nature of the sealant [61].

Mould growth can be prevented or at least reduced by formulating sealants with fungicidal or fungistatic additives [61]. Fungicidal additives kill the mould organisms, while fungistatic additives inhibit their reproduction and growth. Fungicides can only be effective, if they are present on the sealant surface in sufficient concentration to affect the mould growth. In general, fungicide concentrations as low as 10-100 ppm are effective. To ensure that this minimum concentration is maintained on the surface of the sealant, small amounts of fungicide must continuously migrate from the sealant's interior to its surface. If the sealant is frequently sprayed with water, as is the case with sanitary installations, some of the fungicide will be leached out, which limits the effective life of the fungicide.

2.6.1.2 Biodegradation by bacteria

For most civil engineering applications, the resistance of sealants to biodegradation is of decisive importance. Sealants installed in tunnels, sewers or sewage treatment plants are exposed to microbiological attack by aerobic and anaerobic bacteria. The biodeterioration caused by bacteria is primarily a result of enzyme production [58]. One of the first systematic tests on the resistance of sealants to biodegradation was carried out by the German Federal Materials Testing Institute in 1973 [62]. Various natural and synthetic rubber seals and cured-in-place sealants were stored in soil samples and in the sludge tank of a sewage treatment plant. In this test, polyurethane sealants and synthetic rubber seals proved the most resistant to microbiological attack; foamed rubber seals made from blends of natural and synthetic rubbers proved the least resistant. The inadequate resistance of natural rubber was also independently confirmed by a British study [63].

Table 4. Weight loss experienced by various cured-in-place sealants upon storage in waste water over a period of 6 or 12 months (data from [58])

Sealant				Biocide		Weight Loss (%) After	
Number	Type[a]	No. of Parts[b]	Cure System[c]	Addition	Type	6 Months	12 Months
1	PS	2p	Unknown	No			38.4
2	PS	2p	Unknown	No			30.7
3	PS	2p	Unknown	Yes	Unknown		4.5
4	PS	2p	Unknown	Yes	Tar	6.7	
5	PS	2p	Unknown	Yes	Tar	17.9	
6	PU	2p	Unknown	No			0
7	PU	2p	Unknown	No			4.4
8	PU	2p	Unknown	Yes	Tar		0.6
9	PU	2p	Unknown	Yes	Tar		1.5
10	PU	2p	Unknown	Yes	Tar		2.7
11	PU	2p	Unknown	Yes	Tar		4.4
12	PU	2p	Unknown	Yes	Tar		4.7
13	PU	2p	Unknown	Yes	Tar		5.1
14	PU	1p	Unknown	No			2.1
15	Si	1p	BA	No			4.4
16	Si	1p	AC	No			1.3
17	Si	1p	AM	No			0.4
18	Si	1p	AM	Yes	Unknown		0.5

a) Sealant Type: PU: Polyurethane, PS: Polysulfide, Si: Silicone
b) No. of Parts: 1p: one-part, 2p: two-part
c) Cure System: BA: Benzamide, AC: Acetoxy, AM: Amine

Inspection of sewer joints [64] revealed that polysulfide sealants formulated without the use of biocides are specifically prone to microbiological attack by anaerobic bacteria. Wherever polysulfide sealants were exposed to anaerobic waste water, a severe softening occurred which resulted in erosion of the sealant by flowing water. Several polysulfide joints were completely eroded up to a depth of 15 mm within 10 years of service. In a separate study [65], polysulfide sealants, together with several other organic sealants, all formulated without addition of biocides, were found to suffer rapid attack in sewage water. The attack was most pronounced in the approximately 35°C warm water of the digestor tank. Within the first mentioned study additional experiments were carried out by immersing cured sealant samples in untreated waste water of a public sewage treatment plant [64]. After only a few weeks of exposure, polysulfide sealants softened to such an extent that matter was being eroded away. In the same study, polyurethane sealants with and without tar additives as well as silicone sealants showed excellent resistance to microbiological attack. Table 4 shows the weight loss experienced by various cured-in-place sealants upon storage in waste water over a period of 6 or 12 months (data from [64]). In a separate study [66], tar modified polyurethane sealants were found to provide the best overall resistance to bacterial attack.

According to the technical literature of one polymer supplier [67,68], significant improvements in the bacterial resistance of polysulfide sealants can be achieved by use of manganese dioxide as the curing agent and addition of a suitable biocide. These findings were confirmed by a further study of the same polymer supplier [69], which showed that pure polysulfide polymer was inherently resistant to degradation in a bacterial medium, which was known to rapidly degrade certain polysulfide sealants, confirming the hypothesis that polysulfide sealants degrade bacterially through secondary effects. Lead dioxide cured polysulfide sealants in this study showed little resistance to bacterial attack, while some biocides were effective in manganese dioxide cured polysulfide sealants. To confirm the findings of the laboratory study, a number of sealant samples were immersed in the active sludge channel of a sewage treatment plant in England over a period of six years. Surface degradation of polysulfide sealants not designed for this environment occurred within one month of immersion. Degradation advanced rapidly for polysulfide sealants that were either lead dioxide cured or low in polymer content. The study concludes that polysulfide sealants can be formulated for good bacterial resistance by utilising the manganese dioxide curative and a biocide and maintaining a minimum polymer content of 40%.

2. 6.2 Macrobiological Attack

Macrobiological damage to sealants may be caused by plants, animals and man. Plants such as mosses may overgrow sealants. Bacteria settling on the sealant may assist plant growth by biodegrading certain sealant formulation components and supplying nutrients to the plant. For some underground applications, sealants need to resist penetration by tree roots. This is achieved by adding chemicals to the sealant formulation that prevent the growth of roots within the sealant. The bleeding of these additives from the sealant to the environment is undesired and needs to be minimised.

The most prominent attack of insects on sealants is that of termites. Seals of several millimetre depth can be penetrated by these insects within a few days. Termite attacks on sealants can be avoided by adding termite repellent chemicals to the sealant formulation. Sealants may also be attacked by birds. Putties based on vegetable or animal oils are the primary target, since the birds' digestive system is capable of extracting the oils. The high molecular weight polymers contained in modern elastic sealants cannot be digested by them, still birds have been observed to feed even on silicone sealants [70].

Since the early 1980s, the most significant type of macrobiological damage to sealants is human vandalism. Sealants are pulled out of joints, punctured or cut with knives; a phenomenon only too familiar on school premises or urban high-rise buildings.

3 Summary

The author considers three of the individual ageing factors discussed in Part I in this series of papers to have the most prominent effects on the service life of building sealants. The first of these factors is sunlight, which induces the photochemical degradation of a sealant's adhesion to glass or to other substrates transparent to ultraviolet light. Since the adhesion of a sealant to a substrate stems from a relatively low number of chemical bonds, it may be rapidly lost, if a sensitive sealant/substrate interphase is exposed to ultraviolet or visible light. The second factor is cyclic mechanical strain induced in sealants by thermal movement of building components. This cyclic stress may lead to permanent deformations in sealants which exhibit a high plastic deformation component and may subsequently cause them to fail cohesively. It may also cause adhesive failure, if the stresses induced at the sealant/substrate interphase exceed the sealant's bond strength. The third ageing factor is ambient heat, which induces a "post-cure" in most sealants. Since elevated ambient temperatures typically coincide with compression of the sealant in a building joint, the post-cure not only causes an increase in the sealant's modulus, but may also manifest itself as a permanent compression set. Both effects result in higher stresses at the sealant/substrate interphase and, consequently, a higher probability of adhesive failure. Other ageing factors may become more prominent for sealant applications other than above grade building construction. In civil engineering, for instance, hydrolysis by water or microbial attack may be the most important ageing factors. Part II in this series of papers will deal with synergetic effects between ageing factors and attempts at correlating accelerated and natural ageing.

4 Acknowledgements

The author would like to thank Mrs. Virginia O'Neil and Dr. Loren Lower, Dow Corning Corporation, Midland, MI 48686, U.S.A., for critically reading the manuscript prior to publication.

5 References

1.   International Organisation for Standardisation. (1989) *Building Construction - Jointing Products - Sealants - Determination of Adhesion/Cohesion Properties at Variable Temperatures*, ISO, Geneva, Switzerland. ISO 9047.
2.   American Society for Testing and Materials. (1986) *Adhesion and Cohesion of Elastomeric Joint Sealants Under Cyclic Movement (Hockman Cycle)*, ASTM, Philadelphia, U.S.A. ASTM C-719.
3.   American Society for Testing and Materials. (1991) *Effects of Accelerated Weathering on Elastomeric Joint Sealants*, ASTM, Philadelphia, U.S.A. ASTM C-793.

4.   Kockott, D. (1989) Natural and Artificial Weathering of Polymers. *Polymer Degradation and Stability*, Vol. 25, pp. 181-208.

5.   Stolarski, R., Bojkov, R., Bishop, L., Zerofos, C., Staehelin, J. and Zawodny, J. (1992) Measured Trends in Stratospheric Ozone. *Science*, Vol. 256, pp. 342-349.

6.   Gleason, J.F., Bhartia, P.K., Herman, J.R., McPeters, R., Newman, P., Stolarski, R.S., Flynn, L., Labow, G., Larko, D., Seftor, C., Wellemeyer, C., Komhyr, W.D., Miller, A.J. and Planet, W. (1993) Record Low Global Ozone in 1992. *Science*, Vol. 260, pp. 523-526.

7.   Pickett, J.E. (1994) Effect of Stratospheric Ozone Depletion on Terrestrial Ultraviolet Radiation: A Review and Analysis in Relation to Polymer Photodegradation. *Polymer Degradation and Stability*, Vol. 43, pp. 353-362.

8.   Ashton, H.E. (1969) Weathering of Organic Building Materials. *Canadian Building Digest*, Vol. 117, pp. 117/1-4, National Research Council, Division of Building Research, Ottawa, Canada.

9.   Wolf, A.T. (1989) Studies of the Ageing Behaviour of Gun-Grade Building Joint Sealants - The 'State-of-the-Art'. *Polymer Degradation and Stability*, Vol. 23, pp. 135-163.

10. Friberg, G. (1970) Environmental Action on Polymeric Materials (in Swedish). Plastvaerlden, Vol. 3, pp.82-84.

11. Alsleben, G., Mühl, H.-J., Jonas, R. and Wolf, A.T. (1979) Ageing of Elastic Sealants by Sunlight (in German). *Kautschuk + Gummi, Kunststoffe*, Vol. 32, pp. 671-672.

12. Israëli, Y., Lacoste, J., Cavezzan, J. and Lemaire, J. (1993) Photo-Oxidation of Polydimethylsiloxane Oils. Part III: Effect of Dimethylene Groups. *Polymer Degradation and Stability*, Vol. 42, pp. 267-279.

13. Dolezel, B. (1978) *The Durability of Plastics and Rubbers* (In German), C. Hanser Verlag, München and Wien.

14. Spauszus, S. (1975) *Material Science of Glass* (in German), Deutscher Verlag für Grundstoffindustrie, Leipzig.

15. Gjelsvik, T. (1975) *The Effect of the Outdoor Climate upon Materials and Constructions in Facades* (in Norwegian). The Norwegian Building Research Institute, Offprint No. 234.

16. Zimmermann, G. (1968) Thermally Induced Length Changes in Construction Elements (in German). *Deutsche Bauzeitung*, Vol. 25, No. 10, pp. 33-38.

17. Kuenzel, H. and Gertis, K. (1969) Thermally Induced Deformation in Exterior Walls (in German). *Baustein-Zeitung*, Vol. 35, pp. 66-70.

18. Karpati, K.K. (1972) Mechanical Properties of Sealants: I. Behaviour of Silicone Sealants as a Function of Temperature. *Journal of Paint Technology*, Vol. 44, pp. 55-66.

19. Karpati, K.K. (1972) Mechanical Properties of Sealants: II. Behaviour of Silicone Sealants as a Function of Rate of Movement. *Journal of Paint Technology*, Vol. 44, pp. 58-64.

20. Karpati, K.K. (1972) Mechanical Properties of Sealants: III. Performance Testing of Silicone Sealants. *Journal of Paint Technology*, Vol. 44, pp. 75-85.

21. Karpati, K.K. (1973) Mechanical Properties of Sealants: IV. Performance Testing of Two-Part Polysulfide Sealants. *Journal of Paint Technology*, Vol. 45, pp. 49-57.

22. Karpati, K.K. (1979) Development of Test Procedure for Predicting Performance of Sealants, in American Chemical Society Symposium Series No. 113, *Plastic Mortars, Sealants, and Caulking Compounds*, pp. 157-179.

23. Karpati, K.K. (1980) Weathering of Silicone Sealant on Strain-Cycling Exposure Rack. *Adhesive Age*, Vol. 23, pp. 41-47.

24. Karpati, K.K. (1984) Investigation of Factors Influencing the Outdoor Performance of Two-Part Polysulfide Sealants. *Journal of Coating Technology*, Vol. 56, pp. 57-69.

25. Karpati, K.K. (1985) Testing Polysulfide Sealant Deformation on Vices. *Adhesive Age*, Vol. 28, No. 5 (May), pp. 18-22.

26. Karpati, K.K. (1987) Laboratory Fatigue Test of a Two-Part Polysulfide Sealant Correlated to Outdoor Performance. *Durability of Building Materials*, Vol. 5, pp. 35-51.

27. Karpati, K.K. (1988) Exposure Evaluation of Sealants with Low Movement Capability. *Adhesive Age*, Vol. 31, No. 5 (May), pp. 20-23.

28. Karpati, K.K. (1989) Performance of Polyurethane Sealants on a Strain-Cycling Exposure Rack. *Materials and Structures*, Vol. 22, pp. 60-63.

29. Burström, P.G. (1979) *Ageing and Deformation Properties of Building Joint Sealants*, Report TCBM-1002, Division of Building Materials, University of Lund, Lund, Sweden, pp. 34-38.

30. Luck, R.M. and Mendelsohn, M.A. (1982) The Degradation and Outgassing of Polymeric Sealants and Plastics and Their Effect on Solar Collector Efficiency. *Journal of Polymer Preparations, American Chemical Society, Division of Polymeric Chemistry*, Vol. 23, pp. 235-242.

31. Burström, P.G. (1990) European Experiences of Sealants in Service, Correlation to Results from Laboratory Tests, in *Building Sealants: Materials, Properties and Performance*, ASTM STP 1069, (ed. T.F. O'Connor), American Society for Testing and Materials, Philadelphia, pp. 295-302.

32. Beech, J. and Beasley, J. (1994) Further Studies of Cure and Durability of Building Sealants, in *Science and Technology of Building Seals, Sealants, Glazing, and Waterproofing: 3rd Volume*, ASTM STP 1254, (ed. J.C. Myers), American Society for Testing and Materials, Philadelphia, pp. 33-50.

33. International Organisation for Standardisation (1984) *Building Construction - Jointing Products - Sealants - Determination of Tensile Properties*, ISO, Geneva, Switzerland, ISO 8339.

34. Stögbauer, H. and Wolf, A.T. (1990) The Influence of Heat Ageing on One-Part Construction Silicone Sealants, in *Building Sealants: Materials, Properties and Performance*, ASTM STP 1069, (ed. T.F. O'Connor), American Society for Testing and Materials, Philadelphia, pp. 193-208.

35. Yang, A.C.M. (1994) Filler-Induced Softening Effect in Thermally Aged Polydimethylsiloxane Elastomers. *Polymer*, Vol. 35, pp. 3206-3211.

36. Klosowski, J.M. (1989) *Sealants in Construction*, Marcel Dekker, Inc., New York and Basel, pp. 21.

37. American Society for Testing and Materials (1987) *Low-Temperature Flexibility of Latex Sealants After Artificial Weathering*, ASTM, Philadelphia, U.S.A., ASTM C-734

38. Hanheia, P.J., Huang, R.H.E. and Paul, D.B. (1986) Water Immersion of Polysulfide Sealants: I. Effect of Temperature on Swell and Adhesion. *Industrial and Engineering Chemistry Product Research and Devevelopment*, Vol. 25, pp. 328-336.

39. Aubrey, D.W. and Beech, J.C. (1985) The Performance of Joint Sealants Between Porous Surfaces in Wet Conditions. *ASE Conference*, London, Vol. 2, pp. 360-366.

40. Ludwig, B. and Wolf, A.T. (1986) Insulating Glass Sealants - Test and Evaluation Criteria. *Kautschuk + Gummi, Kunststoffe*, Vol. 39, pp. 922-925.

41. Massoth, H. (1987) Water Vapour Transmission Rates and Water Swelling of Insulating Glass Sealants (in German), *Industrial Project Work Thesis*, Fachhochschule Darmstadt, Darmstadt, Germany.

42. Aubrey, D.W. and Beech, J.C. (1989) The Influence of Moisture on Building Joint Sealants, *Building and Environment*, Vol. 24, No.2, pp. 179-190.

43. Boesmans, O. (1990) Comparison of Two Two-Part Insulating Glass Silicone Sealants" (in French), *Industrial Project Work Thesis*, Faculte Polytechnique de Mons, Mons, Belgium.

44. Beech, J. and Mansfield, C. (1990) The Water Resistance of Sealants for Construction, in *Building Sealants: Materials, Properties and Performance*, ASTM STP 1069, (ed. T. O'Connor), American Society for Testing and Materials, Philadelphia, pp. 209-220.

45. Lowe, G.B., Lee, T.C.P., Comyn, J. and Huddersman, K. (1994) Water Durability of Adhesive Bonds Between Glass and Polysulfide Sealants. *International Journal of Adhesion and Adhesives*, Vol. 14, pp. 85-92.

46. Gan, L.M., Ong, H.W., Tan, T.L., Chen, C.H. and Lai, F.N. (1985) Durability of Silicone Sealant Bond on Metal or Glass. *Durability of Building Materials*, Vol. 2, pp. 379-385.

47. Descamps, P., Iker, J. and Wolf, A.T. (1994) Methods of Predicting the Adhesion of Silicone Sealants to Anodised Aluminium, in *Science and Technology of Building Seals, Sealants, Glazing, and Waterproofing: 3rd Volume*, ASTM STP 1254, (ed. J.C. Myers), American Society for Testing and Materials, Philadelphia, pp. 95-106.

48. McCann, K.D. (1990) Sealant Lifetime in an Immersed Environment, *Dow Corning Internal Research Report*, Dow Corning Corporation, Midland, U.S.A.

49. Agger, J.R., Descamps, P., Iker, J. and Tilmant, G. (1991) A Study into the Loss of Adhesion of Silicone Sealant Test Samples, *Dow Corning Internal Research Report*, Dow Corning S.A., Seneffe, Belgium.

50. Iker, J., Descamps, P., Tilmant, G. and Agger, J.R. (1992) Study into the Validity of an Accelerated Test Method for Tensile Adhesion Samples by Immersion into Water at Elevated Temperatures (in French). *Revue Technique du Batiment et des Constructions Industrielles*, Vol. 23, No. 143 (March/April), pp. 18-20.

51. Donaldson, P.E.K. (1994) Hydrothermal Stability of Joints, Using a Silicone Rubber Adhesive, for a Range of Adherents of Interest to Makers of Surgically-Implanted Microelectronic Devices. *International Journal of Adhesion and Adhesives*, Vol. 14, pp. 103-107.

52. Mansfield, C. (1990) Tests for the Water Resistance of Construction Sealants. *Construction and Building Materials*, Vol. 4, pp. 37-42.

53. Keshavara, R. and Tock, R.W. (1993) Modelling of Crosslinking Mechanism When Structural Silicone Sealants are Subjected to Moisture. *Polymer-Plastics Technology and Engineering*, Vol. 32, No. 6, pp. 579-593.

54. Klausmeier, R.E and Jones, W.A. (1961) Microbial Degradation and Plasticisers, in *Developments in Industrial Microbiology*, Vol. 2, Plenum Press, New York.

55. Gross, H. and Becker, H. (1979) Studies of Mould Fungus Growth on Sealants, 1st Bulletin (in German). *Adhäsion*, Vol. 23, pp. 106-112.

56. Gross, H. and Becker, H. (1979) Studies of Mould Fungus Growth on Sealants, 2nd Bulletin (in German). *Adhäsion*, Vol. 23, pp. 338-345.

57. Gross, H. and Becker, H. (1983) Studies of Mould Fungus Growth on Sealants, 3rd Bulletin (in German). *Adhäsion*, Vol. 27, pp. 28-33.

58. Flynn, F.B. and Taylor, W.S. (1989) Biological Degradation, Biocides, in *Encyclopaedia of Polymer Science and Engineering*, Second Edition, Vol. 2, (ed. H.F. Mark, N.M. Bikales, C.G. Overberger, G. Menges, J.I. Kroschwitz), John Wiley & Sons, Inc., New York, pp. 202-219.

59. Sears, J.K. and Touchette, N.W. (1989) Plasticisers, in *Encyclopaedia of Polymer Science and Engineering*, Second Edition, Suppl. Vol., (ed. H.F. Mark, N.M. Bikales, C.G. Overberger, G. Menges, J.I. Kroschwitz), John Wiley & Sons, Inc., New York, pp. 609-647.

60. Darby, J.R. and Sears, J.K. (1975) Fungus Resistance, in *Applied Polymer Science*, (ed. J.K. Craver and R.W. Tess), American Chemical Society, Washington, U.S.A., pp. 610-631.

61. Wolf, A.T. (1989) Mould Fungus Growth on Sanitary Sealants, *Construction and Building Materials*, Vol. 3, pp. 145-151.

62. Kerner-Gang, W. (1973) Resistance of Elastomeric Sealing Materials to Mirco-Organisms (in German), *Material und Organismen*, Vol. 8, pp. 17-37.

63. Anonymous (1978) Biodeterioration of Rubber Sealing Rings in Water and Sewage Pipelines, *Notes on Water Research*, No. 18 (November), Water Research Centre, London.

64. Schremmer, H. (1981) Two-Component Sealants for Sewage Pipes (Bilingual: German/English). *Tunnel*, Vol. 16, No. 1, pp. 42-56.

65. Appleton, B. (1973) Biodegradation: Coming Apart at the Seals, *New Civil Engineer*, Vol. 13, No. 12 (December), pp. 6-9.

66. Engelmann, H. (1986) Two-Part Sealants for Sewage Treatment Plants (in German), *Kunststoffe*, Vol. 76, pp. 93-96.

67. Anonymous (1986) *Biodegradation Studies: Status Report, September 1982*, Thiokol Chemicals Limited, Coventry, United Kingdom.

68. Anonymous (1986) *Compounding Polysulfides for Resistance to Biodegradation*, Thiokol Chemicals Limited, Coventry, United Kingdom.

69. Lee, T.C.P. (1988) Formulating Sealants for Long-Term Water Immersion - A Six Year Field Trial, in RILEM International Symposium Building Joint Sealants, Boras, Sweden, 18-19 May 1988, (reviewed by A.T. Wolf), *Kautschuk + Gummi, Kunststoffe*, Vol. 41, pp. 1251-1258.

70. Leempoel, P. (1994), private communication, Dow Corning S.A., Seneffe, Belgium.

7 CONTRIBUTION TO VALIDATION OF LABORATORY TEST METHODS FOR PREDICTION OF THE DURABILITY OF BUILDING JOINT SEALANTS

H. BOLTE and T. BOETTGER
University of Leipzig, Germany

Abstract

At present, in our research group at the University of Leipzig the following tests are being carried out on selected elastomeric sealants (Polysulphide, Polyurethane, Silicone, and for comparing EPDM and Polychloroprene):

• Alternate storage with fluorenscent UV lamps
• Alternate storage with xenon arc lamps
• Thermal storage
• Open air weathering
• Room storage

We characterise the change in materials properties through mechanical test methods with tensile strength and elongation at break.

The intermediate evaluation of the results shows, that the similarity of the curves of artificial ageing and the open air weathering is clearly visible. A precise assessment of the time-compression factor is not possible at the present time.

The tests on the materials will be continued, possibly with different parameters, until virtual agreement with natural weathering is produce on all materials groups, at least with an accelerated ageing method.

Keywords: Elastomeric sealants, accelerated ageing, fluorescence lamp, xenon arc, heat, tensile strength, elongation at break.

1 Introduction

The problems of long-term testing of sealants arise due to the multiplicity of influencing factors. By comparison to the extent of use, a high level of testing and time is required, which has not been possible to accomplish with in the past. For

Durability of Building Sealants. Edited by J.C. Beech and A.T. Wolf. © RILEM.
Published by E & FN Spon, 2–6 Boundary Row, London SE1 8HN, UK. ISBN 0 419 21070 9.

this reason, the long-term stability is usually assessed on the basis of practical experience [11]. In the interest of the quality of the sealants, it is thus necessary to develop a suitable test method to a stage where it can be standardized. For this purpose, the test conditions must be simplified and precisized coupled with a reduction in the period of testing by restricting the parameters for artificial ageing.

2 Experimental

2.1 Tested materials
The tested elastomeric sealants (polysulphide rubber, silicone rubber, polyurethane, compared with polychloroprene rubber and EPDM rubber) are listed in Table 1.

Table 1. List of the tested elastomeric sealants

Nr.	Raw material basis	Reaction system	Charge-Nr.
A 5	Silicone rubber	1 P / neutral / Oxime	3/6/92
A 6	Silicone rubber	1 P / basic / Amine-Oxime	2221113/92
A 7	Silicone rubber	1 P / neutral / Oxime	8690/92
A 8	Silicone rubber	1 P / neutral / Oxime	25/6/92
A 17	Silicone rubber	1 P / neutral / Alkoxy-Titanium	3/6/92
A .18	Silicone rubber	1 P / neutral / Benzamideo-Titanium	2/6/92
A 19	Silicone rubber	1 P / neutral / Benzamide	27/5/92
A 21	Silicone rubber	1 P / acid / Acetate	13/05/92
B 9	Polysulphide rubber	2 P / Manganese dioxide	677/8/89
B 10	Polysulphide rubber	2 P / Manganese dioxide	39/04/872 260/8/89
B 12	Polysulphide rubber	2 P / Manganese dioxide	54/12/88 19/03/90
B 22	Polysulphide rubber	2 P / Lead dioxide	010721/92
B 28	Polysulphide rubber	prefabricated	12/91/654
C 23	Polyurethane	1 P	05113451/92
D 24	Polychloroprene rubber	Hot vulcanization	4/79
E 25	EPDM rubber	Hot vulcanization	9/2/82

Where these were unavailable in a prefabricated manner as foil or strip, they were applied with a template between August and October 1992. With two-component sealants, the components were individually and subsequently thoroughly mixed together. Strips were produced with a width of 80 mm, 2-3 mm thick and about 0.9m in length. These were deposited on a horizontal surface and from the following day onwards were suspended for storage. The polyethylene foil was taken off after about 8 weeks. In accordance with the amount available, a strip of 7 to 20 m was produced from every type of material. This is equivalent from 450 to 1300 pieces of rod S2 according to DIN 53504 resp. ISO 37, 1977 [9] per type of material. With this considerable quantity of uniform samples, it is possible to supply a test schedule over several years with comparable material.

2.2 Test criteria

All samples are stored prior to testing at a standard climate 23/50 (DIN 50014 [6]) until the weight was constant. Weighing was precise to 0.01 g. After this, from the 45 mm strip sections, 3 butt strips S2 were punched out (see Fig. 1). The tensile strength and the elongation at break was tested according to DIN 53504 resp. ISO 37, 1977 [9]. This initially eliminates the effects of the surface, since on the majority of sealants these do not have a negative long-term effect [3].

2.3 Test series

2.3.1. Alternate storage with fluorescent UV lamps

8 h UV at +80°C / 4 h CON at +50°C
UVCON device Atlas Co. with UVA 340 lamps
500, 1000, 1500, 2000 .....6000 h

The cycle of 8 h UV radiation/4 h condensation was chosen from ASTM G 53-88 [1] and ASTM D 4329-92 [2]. The black standard temperatures +80°C resp. +50°C were specified on the basis of our own experience.
To achieve uniform exposure of the samples in the UVCON device, horizontal rotation was carried out every 3-4 days, as well as vertical rotation every 3 weeks. Rotation of the lamps took place after 400 ...450 h UV. Temperature inspection produced the following relatively slight fluctuations :
+50°C : 50.3 ... 51.4°C
+80°C : 78.7 ... 82.7 °C

2.3.2. Alternate strorage with xenon arc lamp

3 x (3 d +80°C, 1 d H_2O, 2 d +80°C, 1 d H_2O)
21 d UV with SUNTEST CPS unit Heraeus Co. with xenon arc lamp NXE 1500
1000, 2000, 3000 .....6000 h

The 3 weeks cycle heat/water was obtained from DIN EN 28339 [5], the temperature increased from +70°C to +80°C and supplemented with 3 weeks UV irradiation. The rearrangement took place manually between the ovens, water tanks and SUNTEST units. The heat storage took place according to Section 2.3.3 with 10 changes of air per hour. The water storage took place according to DIN 53495 [8] suspended in distilled water. The irradiation took place according to ISO 11431 without the effect of water [13]. On the surface of the sample, the radiation intensity covers a range from 300 to 800 nm 1000 + 200 W / m². The black standard temperature was measured in the unit with the temperature sensor supplied by the manufacturer over a period of 14 days around 60 times, and compared with the room temperature. It was evident that the black standard temperature was about 30 K above the room temperature measured at the same 2 m distance and could be favourably influenced by installing the units in a very large room (100 m²). The room temperature fluctuates seasonally between +18°C and +34°C. Through the simultaneous use of two SUNTEST units with alternately offset emitter exchange and

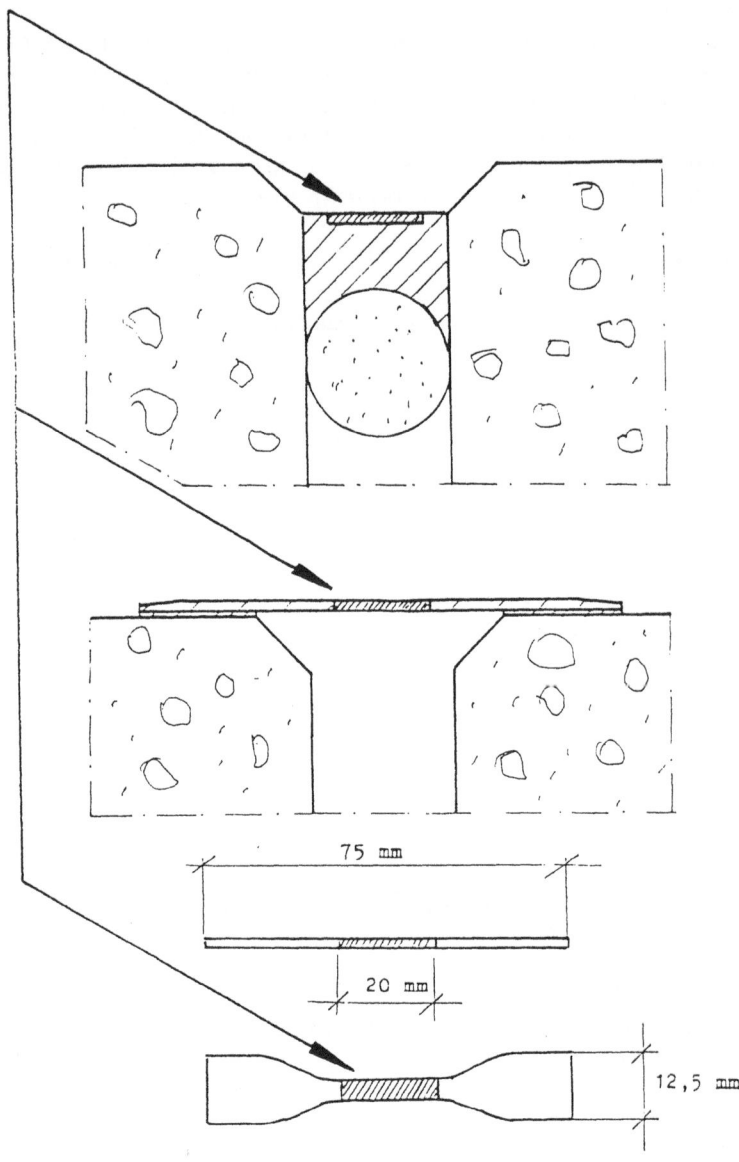

Fig. 1 : Relationship between joint cross-section and test
item shape (DIN 53504 / ISO 37-1977)

systematic crosswise exchange of the strips in the unit and between the units, uniform exposure was achieved.

2.3.3. Thermal storage

+80°C / 10 changes of air per hour
Individual storage in ovens
1000, 2000, 3000 .....6000 h

+100°C / 10 changes of air per hour
Individual storage in ovens
500, 1000, 1500, 2000 .....3500 h

Storage took place suspended in ovens with natural ventilation in accordance with DIN 53508 [10]. The degree of ventilation was determined in accordance with DIN VDE 0304, Part 24-1, by determining the additional power consumption [4]. Strip sections of 45 mm in length, were spaced and suspended from the walls in the ovens, so that the centre of the samples were level in height in the middle of the cabinet. The cabinet temperature was also measured at this point. In each of the small ovens, only samples of one material were located (individual storage).

2.3.4 Open air weathering

Storage took place in accordance with DIN 53386-A [7] at the site in Leipzig, in a temperate industrial climate. The strip-shaped material was fastened on the surface pointing at an angle of 45° southwards, so that the decisive centre section for the test had no contact with the surface.

2.3.5. Room storage

Storage took place in the dark, at room temperature, in accordance with DIN 53386-A [7], an air temperature range of 13°C to 22°C applies. The strip-shaped material was suspended for storage. In parallel to the open air weathering, the mechanical charateristics were determined.

3 **Results and Discussion**

The previous results from 2 typical types of materials were shown in Fig. 2 - 13 :

- Despite precise work, the measured data deviates. The deviations were mainly caused by irregularities in the sample material. One-component systems deviate in some cases, far more than well mixed two-component systems. Even so, by means of storage for 6000 h and testing at intervals of 500 h, it was possible to clearly record the important medium and late curves of artificial ageing.
- The similarity of the curves of artificial ageing and the open air weathering is clearly visible.

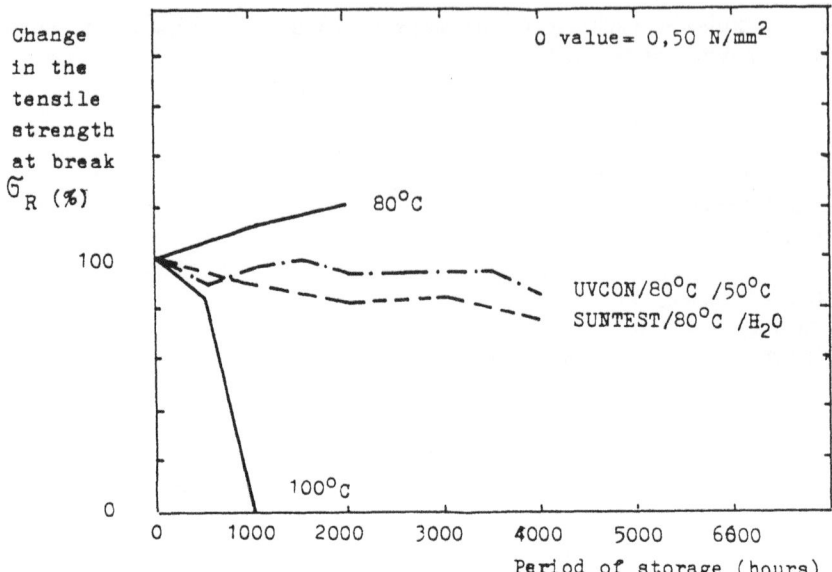

Fig. 2 and 3 : Material B 10, artificial ageing

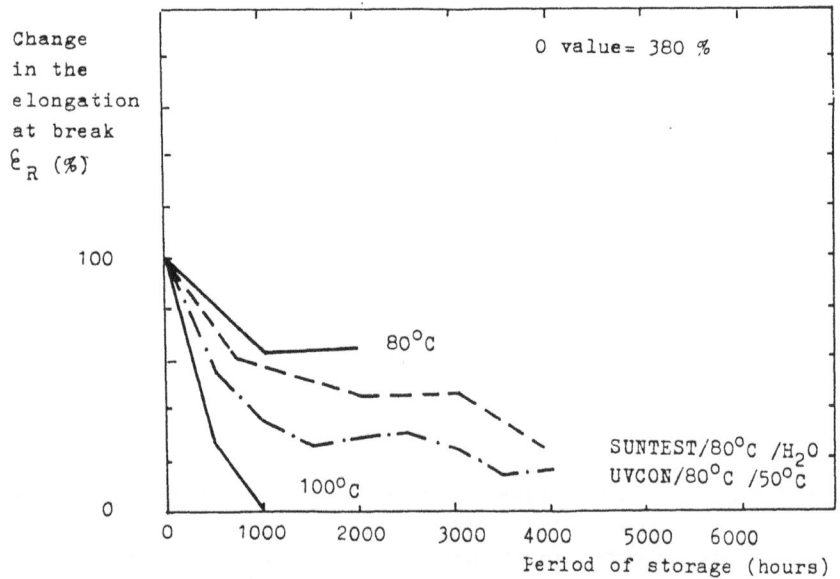

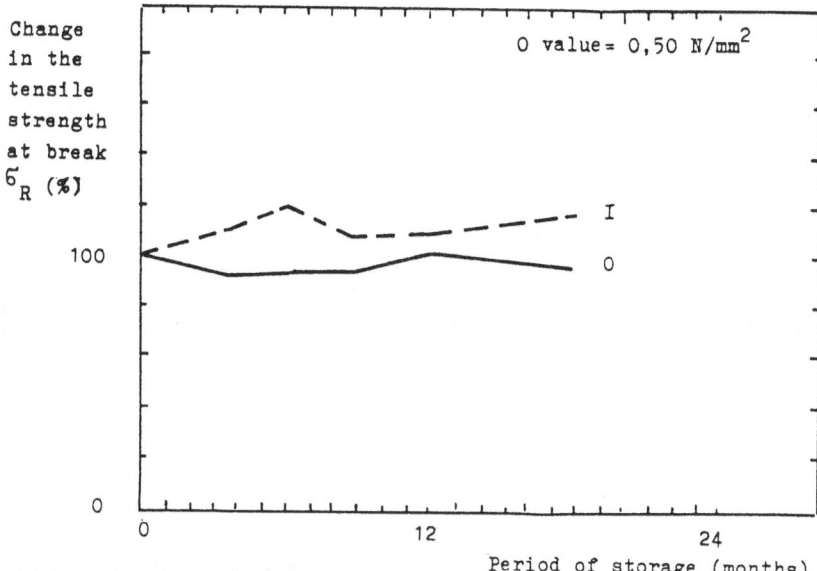

Fig. 4 and 5 : Material B 10, open-air weathering (O)
and indoor storage (I)

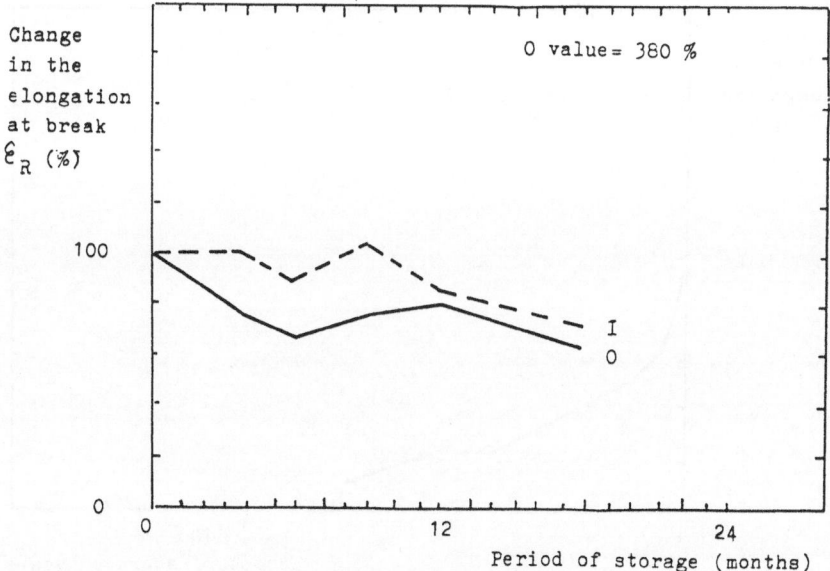

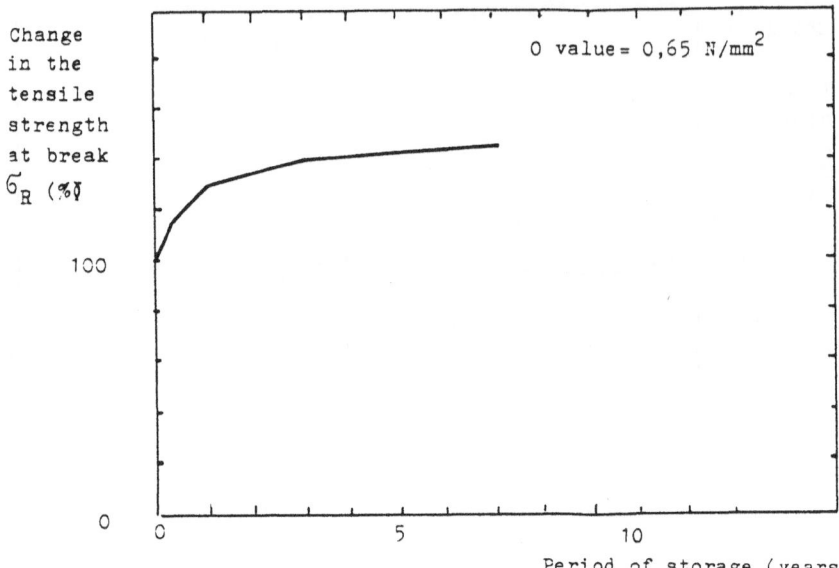

Fig. 6 and 7 : Material B 10 (old batch), open-air
                        weathering

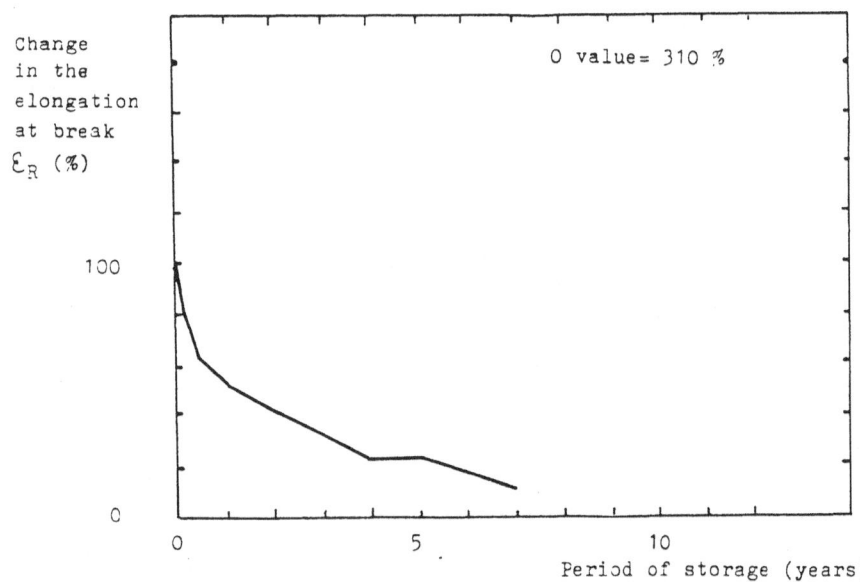

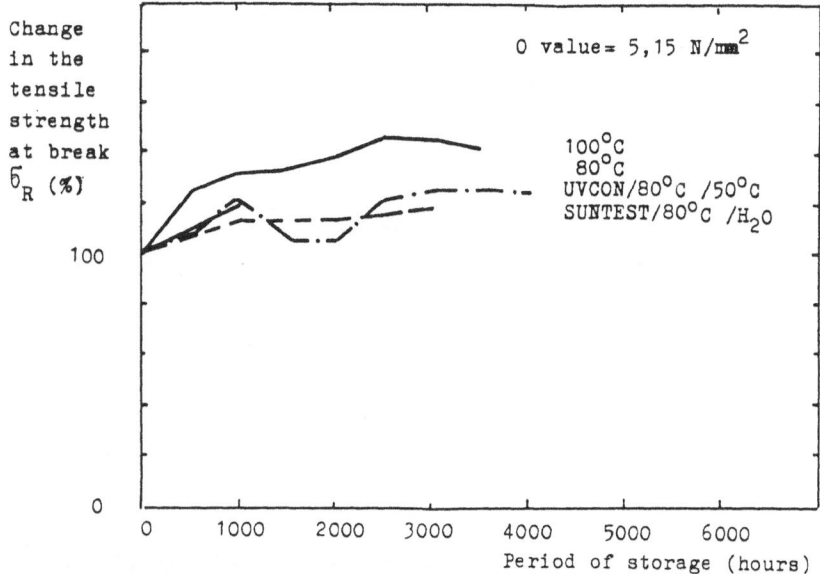

Fig. 8 and 9 : Material E 25, artificial ageing

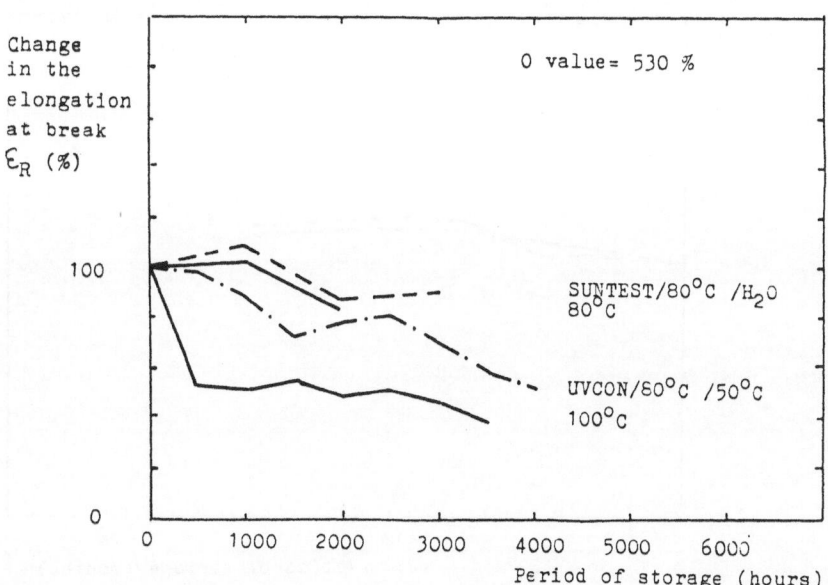

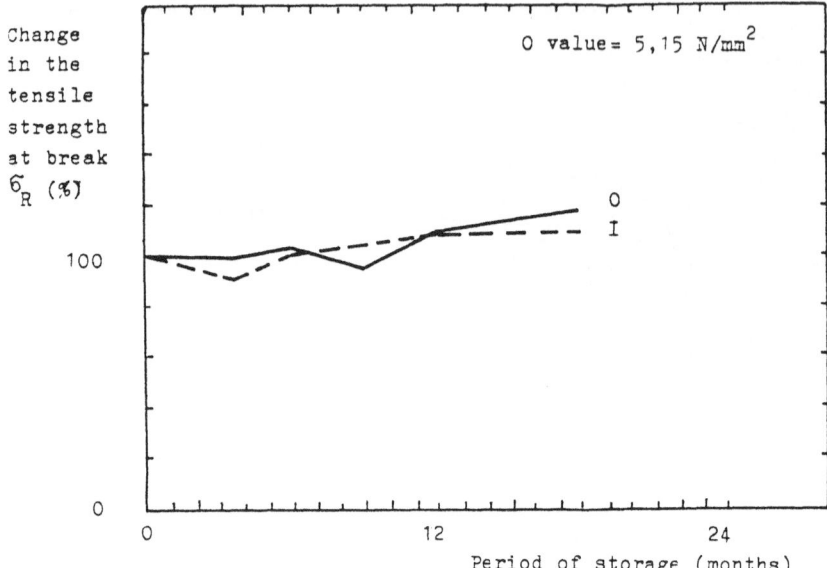

Fig. 10 and 11 : Material E 25, open-air weathering (0)
and indoor storage (1)

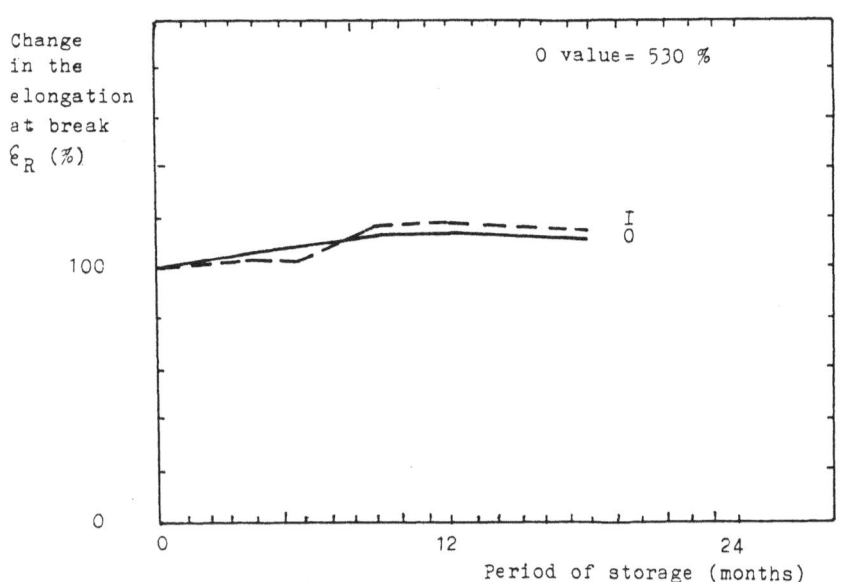

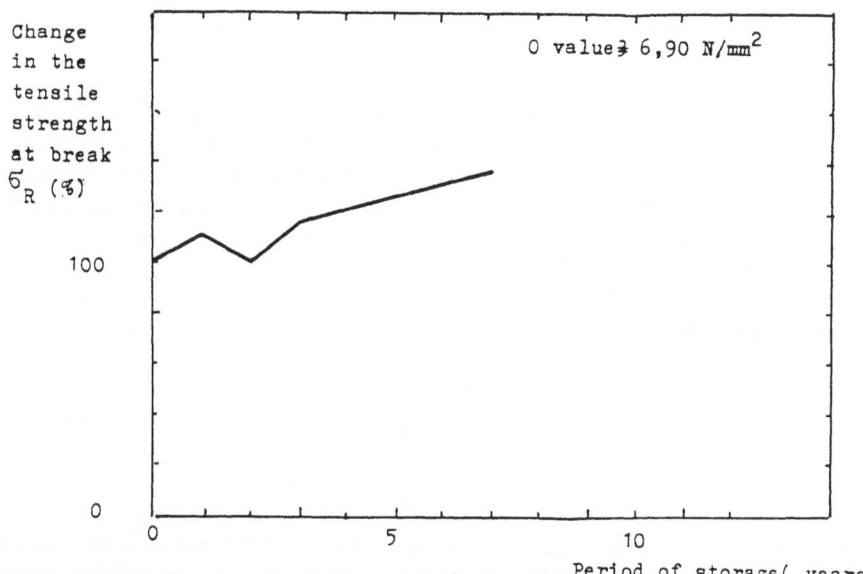

Change
in the
tensile
strength
at break
σ_R (%)

O value ≠ 6,90 N/mm^2

100

0

0   5   10

Period of storage(years)

Fig. 12 and 13 : Material E 25 (old batch), open-air
weathering

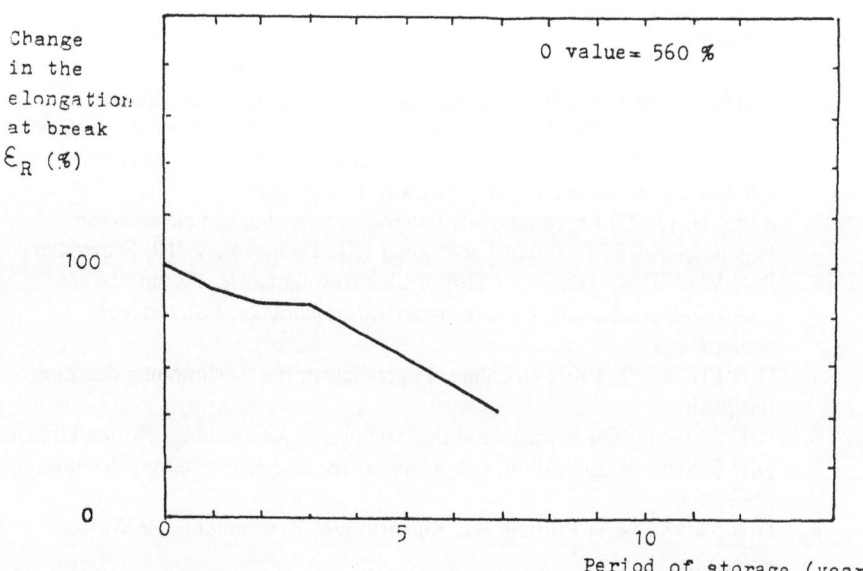

Change
in the
elongation
at break
ε_R (%)

O value = 560 %

100

0

0   5   10

Period of storage (years

A precise assessment of the time-compression factor is not possible at the present time.

- Open air weathering and room storage after 18 months produce virtually the same data for most materials. Even on both the foils that had been rolled up and stored for about 10 years, the same applies. This clearly shows the major importance of atmospheric oxygen, which had a considerable effect on the open air weathering during the initial phase.

4 Conclusions

- The tests on the strips will be continued, possibly with different parameters, until virtual agreement with natural weathering is produced on all material groups, at least with an accelerated ageing method.

- Confirmatory trials will be subsequently carried out in accordance with DIN EN 28339 [5] on joint models, which have also been produced in Leipzig in 1992, and have been stored partly indoors and partly in the open. In addition, the effect of the mechanical strain is to be verified. For this purpose, sealants are to be artificially weathered, e.g. for 500, 1000, 1500, 2000 and more hours, and are to be simultaneously subjected to constant strain or to stress cycles before and after each individual period of time, in accordance with ISO 9047 [12].

5 References

1. ASTM G 53-88 (1988) Operating light- and water-exposure apparatus (Fluorescent UV-Condensation Type) for exposure of nonmetallic materials.
2. ASTM D 4329-92 (1992) Operating light and water apparatus (Fluorescent UV and Condensation Type) for exposure of plastics.
3. Bolte, H. (1992) Ergebnisse aus Langzeituntersuchungen elastomerer Fugendichtstoffe, IBK-Bau-Fachtagung 151, Hannover, 9./10. September.
4. DIN VDE 0304, Teil 24 - 1 (1990) Elektroisolierstoffe, thermische Langzeiteigenschaften, Einzelkammerwärmeschränke, Entwurf vom September.
5. DIN EN 28339 (1991) Hochbau, Fugendichtstoffe, Bestimmung der Zug-festigkeit.
6. DIN 50014 (1985) Klimate und ihre technische Anwendung, Normalklimate.
7. DIN 53386 (1982) Prüfung von Kunststoffen und Elastomeren, Bewitterung im Freien.
8. DIN 53495 (1984) Prüfung von Kunststoffen, Bestimmung der Wasser-aufnahme.
9. DIN 53504 (1985) Prüfung von Kautschuk und Elastomeren, Bestimmung von Reißfestigkeit, Zugfestigkeit, Reißdehnung und Spannungswerten im Zug-versuch.
10. DIN 53508 (1977) Prüfung von Elastomeren, künstliche Alterung.

11. Grunau, E.B., Esser, S., Jahn, K.J. (1992) Fugen - Auslegung und Abdichtung, expert verlag, Ehningen.
12. ISO 9047 (1989) Building construction - Sealants - Determination of adhesion/cohesion properties at variable temperatures.
13. ISO 11431 (1993) Building construction - Sealants- Determination of adhesion/cohesion properties after exposure to artificial light through glass.

8 DURABILITY OF RESEALED BUILDING JOINTS

A. PAGLIUCA and A.R. HUTCHINSON
Joining Technology Research Centre,
Oxford Brookes University, Oxford, UK

Abstract
Resealing of joints in a typical building structure needs to take place a number of times within the life-time of the building . The resealing process is more difficult than sealing joints for the first time and among the most significant difficulties are the removal of old sealant, and the likely presence of residues left in the joint. This paper reviews the fundamental concepts of adhesion and outlines the use of surface analysis techniques, in particular surface energy measurements, to predict the adhesion and durability performance of resealed joints. Experimental tensile adhesion joints were prepared with a range of contaminated substrate surfaces to simulate typical surfaces found in commercial resealing. Joints were tested before and after QUV accelerated weathering and trends in performance compared favourably with those predicted from surface energy measurements.

Keywords: Buildings; Durability; Extension at failure; Failure mode; Resealing; Surface analysis; Surface energy.

1 Introduction

A substantial proportion of the cost of building maintenance is attributable to resealing of the external envelope, a task which is significantly more difficult than sealant installation during new build. In 1990 approximately 100,000km of building joints were resealed in the U.K. at an installed cost in excess of £500 million. Resealing of joints in a typical building structure needs to take place a number of times within the life-time of the building. A well-sealed joint has a life expectancy less than that of a typical building, perhaps 25 years compared to 60 years or more; a poorly sealed joint has a life expectancy frequently less than 10 years. Indeed, a survey of resealing activity carried out in the U.K. in 1990[1] revealed that 55% of building joint seals had failed within 10 years service; only 15 % had lasted for more than 20 years.

Durability of Building Sealants. Edited by J.C. Beech and A.T. Wolf. © RILEM.
Published by E & FN Spon, 2–6 Boundary Row, London SE1 8HN, UK. ISBN 0 419 21070 9.

It was also noted that adhesion failures, that is failures between sealant and substrate, predominated. Interestingly, the findings of this survey mirrored those of a study carried out in Japan in 1984[1]. The resealing process is more difficult than sealing for the first time, but many building joints are not designed to be resealed despite this obvious need. Among the most significant difficulties are the removal of old sealant and the likely presence of residues left in the joint. This aspect of obtaining adhesion to contaminated surfaces represents the main theme of this paper.

The problem of resealing joints in building structures was addressed comprehensively during the early 1990s by the RESEAL project, run within the UK DOE/EPSRC LINK CMR programme. Participants in the project included Oxford Brookes University (lead partner); Taywood Engineering Ltd.; a consortium of sealant manufacturers consisting of Adshead Ratcliffe Ltd., Evode Ltd., Fosroc-Expandite Ltd., Morton International Ltd. and Sika Ltd.; and advice from the Association of Sealant Applicators and the Building Research Establishment. The purpose of the RESEAL project was to study the performance of sealant materials applied to contaminated substrates and thus provide objective guidelines for the resealing of building structures. The two primary threads of the project were:

- Scientific analysis of sealed and resealed joint performance, including aspects of curing, testing, material characterisation and surface analysis.
- Development of 'Resealing of Buildings: A Guide to Good Practice'[1], based upon the experience of the industrial participants and backed up by the results of the experimental work.

This paper contains some of the experimental findings arising from the RESEAL project. In particular it is concerned with the use of surface analysis techniques such as surface energy measurements, to predict the adhesion and durability performance of tensile adhesion joints prepared with sealants and simulated contaminated substrate surfaces typical of those encountered in commercial resealing.

2 Fundamental concepts

2.1 Background
A construction sealant's primary function is to make the joints in a building structure weathertight. To achieve this it must adhere sufficiently well to the relevant substrates and exhibit adequate cohesive strength and resistance to degrading factors. The durability of any resealed joint is dependent on how these degrading factors are resisted, and adhesion maintained, during the life of the building. The typical degrading factors involved include[2]: cyclic joint movement; U.V.; heat; moisture; wind pressure; oxygen/ozone. Studies have shown that adhesion failures predominate in typical joint configurations in buildings.

It is clear that good long-term adhesion between a sealant and substrate is essential for a durable resealed joint to be formed. However, in resealed joints the development of good adhesion is often compromised by the presence of residue sealant contamination from the original seal[1]. In the resealing of building joints it is desirable to remove all traces of the original sealant before applying the reseal. However in commercial operations the quality of this removal procedure can vary and residue sealant can often remain in the joints, especially in areas of difficult access.

2.2 Adhesion and Bond Durability

The main objective of the work described in this paper was to use current theories of adhesion to predict the bondabilty and durability of various sealant/substrate, and reseal system/contaminated substrate, combinations. To obtain good adhesion it is essential to obtain intimate molecular contact between the sealant and substrate surface. Attaining this ideal situation relies upon a number of factors and can be linked through four main theories of adhesion:

- •Mechanical Interlocking
- •Adsorption Theory
- •Electronic Theory
- •Diffusion Theory

The theories of most relevance to sealant technology are mechanical interlocking and adsorption. The theory of mechanical interlocking is instinctive in that it proposes that the sealant keys into the irregularities of the substrate, thereby creating an adequate bond. This theory is very much substrate dependent and cannot be universally applied. However, the adsorption theory can be very widely used to predict bond interactions[3,4].

Any solid material has an intrinsic cohesive mechanical strength related to the various forces of attraction between the particles of which it comprises. The adsorption theory suggests that given sufficient intermolecular contact at the interface, two materials will adhere because of the action of the same forces of attraction between atoms in the two surfaces. These forces can be divided into primary bonds(ionic, covalent and metallic) and secondary bonds (acid-base interactions and several van der Waals forces).

Table 1 Typical Bond Types

Bond Type	Bond Energy(kJmol^{-1})	Bond Distance(nm)
Primary:		
Ionic	590-1050	≈ 0.105
Covalent	63-710	≈ 0.154
Metallic	113-347	≈ 0.090
Secondary:		
Hydrogen bonds	≈ 10-42	≈ 0.160-0.185
Dip-induced dip	< 2	≈ 0.195-0.220
Dispersion forces	0.08-42	≈ 0.165-0.220

It can be seen from table 1 that these forces are only effective over comparatively short atomic distances - at most a few Angstrom units. Therefore in order for these forces to contribute towards adhesion there must be close and intimate contact between the sealant and substrate at the interface[5]. This fact gives rise to the importance of

concepts involving wetting, surface-free energies and contact angles in adhesion.
The ability of a liquid (such as an uncured sealant) to wet out a solid surface is related
to the relative specific surface free energies of the two surfaces[6]. The surface free
energy of a liquid can be measured directly by measuring its surface tension. However
the surface energy of a solid cannot be measured directly but can be calculated from
measuring the contact angles made by liquids (of known surface energy) on the solid
surface (Fig. 1.).

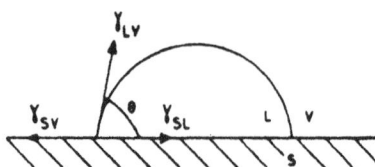

The Young equation resolves the surface tension vectors at the point of contact of a
liquid drop on a solid surface:

$$\gamma_{LV} \, Cos\theta \; = \; \gamma_{SV} - \gamma_{LS} \tag{1}$$

γ_{LV}	=	surface tension of liquid/vapour interface
γ_{LS}	=	surface tension of solid/liquid interface
γ_{SV}	=	surface tension of solid/vapour interface
θ	=	equilibrium contact angle

The Dupre equation considers the work required to separate a layer of liquid from a
surface, reversibly

$$W_A \; = \; \gamma_{LV} + \gamma_{SV} - \gamma_{LS} \tag{2}$$

Combining the two equations gives the Young-Dupre equation (ignoring the spreading
pressure πe which has been shown to be negligible for contact angles greater than
10°[7])

$$W_A \; = \; \gamma_{LV} \, (1 + Cos\theta) \tag{3}$$

W_A is the work of adhesion, the change in free energy per unit of interface.

It is common to sum all the energy available from the various forces of attraction at
a surface and refer to it as the surface free energy γ of the material. Hence, the
reversible thermodynamic work of adhesion (W_A) can be expressed as :

$$W_A \; = \; \gamma_L + \gamma_S - \gamma_{LS} \tag{4}$$

γ_{LS}, the interfacial free energy, represents a disparate collection of energies (both polar and non-polar) and can be simplified by dividing it into two components: γ_{LS}^{d} representing 'dispersion' forces and γ_{LS}^{p} representing 'polar' forces. There is justification for deriving γ_{LS}^{d} as the geometric mean of γ_{l}^{d} and γ_{s}^{d} when dispersion forces alone are present[8]:

$$\gamma_{LS} = \gamma_{L} + \gamma_{S} - 2(\gamma_{L}^{d}\gamma_{S}^{d})^{\frac{1}{2}} \tag{5}$$

Some workers, such as Kaelble[9] , have proposed that a similar step can be taken for the γ^{p} components of interfacial free energy. Combining the two:

$$\gamma_{LS} = \gamma_{L} + \gamma_{S} - 2(\gamma_{L}^{d}\gamma_{S}^{d})^{\frac{1}{2}} - 2(\gamma_{L}^{p}\gamma_{S}^{p})^{\frac{1}{2}} \tag{6}$$

Substituting (6) into (4) provides a geometric mean equation to calculate γ_{s}^{p} and γ_{s}^{d}.

$$W_{A} = 2[(\gamma_{L}^{d}\gamma_{S}^{d})^{\frac{1}{2}} + (\gamma_{L}^{p}\gamma_{S}^{p})^{\frac{1}{2}}] \tag{7}$$

Combining (7) with (3) gives :

$$Cos\theta = (2[(\gamma_{L}^{d}\gamma_{S}^{d})^{\frac{1}{2}} + (\gamma_{L}^{p}\gamma_{S}^{p})^{\frac{1}{2}}] \div \gamma_{L}) - 1 \tag{8}$$

So by using two (or more) liquids ,L, each with known values of γ_{L}, γ_{L}^{d}, γ_{L}^{p}, the two unknown quantities γ_{s}^{p}, γ_{s}^{d} - the solid surface energy components which are not susceptible to direct measurement - may be calculated.

This thermodynamic approach allows the surface energy of a solid surface (including its relative polar/dispersive nature) to be calculated. If the surface energy of a sealant is also measured, an indication of the likely adhesion and potential bond stability between the two can be obtained. The terms Surface free energy and Surface tension of a liquid are interchangeable, being numerically identical but dimensionally different. For an uncured sealant to effectively wet out a substrate it must possess a lower surface tension. The higher the surface energy of a substrate surface, the greater the chance of achieving good wetting with a sealant.

The picture is complicated by the relative amounts of polar/dispersive component of surface energy. A high polar component of surface energy is desirable for promoting initial adhesion. However, it could potentially be detrimental for long-term durability because polar bonds can aid the ingress of moisture, a main agent of degradation[10].

2.2.1 Real Surfaces

This research investigated the surface energy characteristics of several contaminated substrate surfaces and sealant materials to assess the likely adhesion and durability of typical resealed joints. The literature value as measured in vacuo for the surface energy of a material such as anodized aluminium is very high[11]. However, the anodized aluminium used in construction applications will have a much lower surface energy because of different anodizing requirements and the also through the lack of 'laboratory clean' conditions[12]. In resealing situations the surface characteristics will be influenced by the likely presence of residue contamination. This could interfere with the formation of durable bonds and this represents the main focus for investigation.

2.2.2 Primers

Primers are widely used in promoting adhesion to difficult surfaces such as porous substrates and most polymeric low energy surfaces. Low surface tension primers possess inherently good wetting and their main function is to improve the substrate surface characteristics by the introduction of chemical groups which can promote adhesion.

The type of primer used in reseal operations is commonly determined by the type of substrate present, as for new build, despite the presence residue sealant contamination left behind after removal of the original sealant. This is not generally considered when primers are chosen and could have an adverse effect on adhesion and durability of resealed joints. This factor was also investigated.

3 Experimental

3.1 Materials
3.1.1 Sealants and Primers

The sealants chosen see (table 2) for investigation in this research mirrored the trend in UK construction towards the use of higher quality replacement sealants as opposed to 'mastics'. It was felt that representative commercial sealants should be used in the research in order to make the results more easily transferable to industry. This was especially relevant since the research was part-funded by the sealants industry. The alternative of using model formulations would have aided the fundamental scientific approach of the research, but the results would not then have been directly relevant to current industrial practice.

Table 2 Sealants evaluated

Sealants
1-part polysulphide(1ps)
2-part polysulphide(2ps)
1-part polyurethane(1pu)
2-part polyurethane(2pu)
benzamide cure silicone(sil b)
oxime cure silicone(sil o)

The sealants chosen were sourced from within the RESEAL projects' industrial sponsors group. The primers used in the research work were those recommended by the sealant manufacturers for their specific sealants, which were commercially available.

3.1.2 Residue Contamination Used

In choosing the sealants to be used as residue contaminants, it was decided to include both old non-curing mastics and modern curing sealants, as encountered in commercial resealing.

Table 3 Contaminant sealants used

Curing sealants	Polysulphide
	Silicone
	Solvented acrylic
	Polyurethane
Non-curing sealants	GP Mastic
	Bitumen
	Butyl Rubber

3.1.3 Substrates

The substrates chosen for investigation were representative of those encountered in resealing. They were :
• Anodised aluminium (sulphuric acid anodised to BS1615)
• Powder-coated aluminium (polyester coated to BS6496)
• Pre-cast concrete

3.2 Surfaces Investigated
The variability found in removal of existing sealants prior to resealing in commercial operations meant that a three different levels of contamination were investigated:
• Zero sealant contamination - efficient cleaning which has generated a fresh substrate surface, eg. cutting back or grinding a concrete panel
• Thin layer contamination(< 100 microns of residue sealant) - *most* of the sealant mechanically removed followed by solvent cleaning simulating a good commercial cleaning operation (Fig. 2)
• Thick layer contamination - a sealant surface (> 0.5mm of sealant residue). Poor/minimal mechanical cleaning which in effect has left a sealant surface, simulating a bad commercial cleaning operation (Fig.3).
The preparation procedure used to provide these surfaces is shown in Appendix 1.

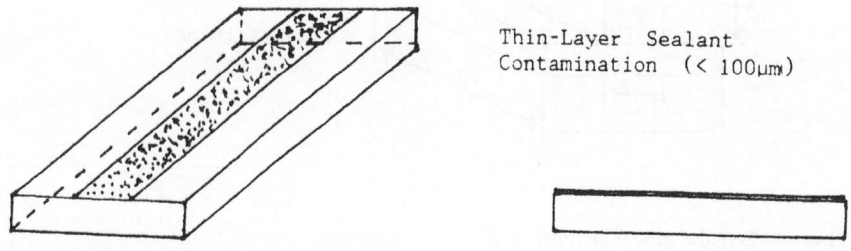

Thin-Layer Sealant
Contamination (< 100μm)

Fig. 2. Thin-layer contaminated substrate

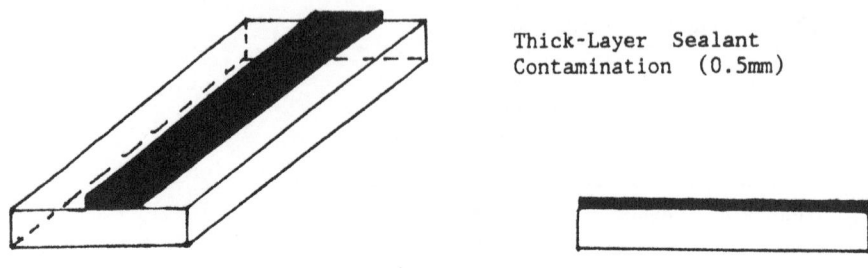

Thick-Layer Sealant
Contamination (0.5mm)

Fig. 3. Thick-layer contaminated substrate

3.3 Preparation and Testing of Tensile Adhesion Test Joints

Tensile adhesion testing of joints made with clean, thin-layer contaminated and thick-layer contaminated substrates was performed. Figure 4 shows the geometry of the test joint used which was based upon an existing BS3712 pt.4 test geometry. Specimens were prepared and cured for 6 weeks outside prior to testing, a curing period and regime which was shown to result in a satisfactory level of cure for all of the resealing sealant materials[13]. Figure 5 shows the detailed nature of the 'hybrid' tensile specimens prepared with the contaminated substrates. These specimens consisted of a 'reference' end, which was anodised aluminium in all cases, and the 'reseal' end.

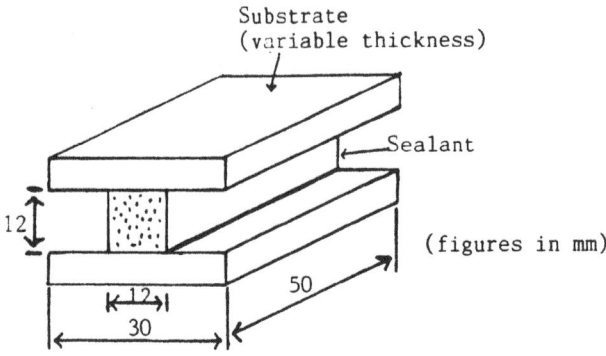

Fig. 4. Tensile adhesion joint

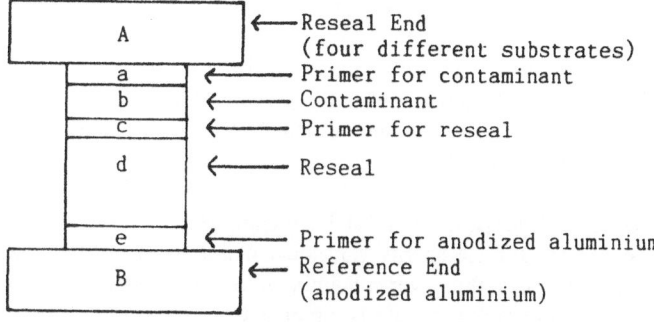

Fig. 5. 'Contaminated test joint'

3.4 Durability Testing

Long-term durability of the test joints was assessed by accelerated ageing. The accelerated ageing regime adopted consisted of 2000 hours exposure in a Q-Panel QUV weatherometer. This was equipped with UVA 340 ultra violet lamps. The joints were subjected to a 4 hour cycle in the QUV weatherometer over the 12 week period of:

    2 hour - 50^0C black panel temperature (UV on)
    2 hour - 40^0C black panel temperature (UV off), condensing humidity

Q-Panel weatherometers were used because several studies have shown that QUV ageing with UVA lamps provide good correlation with natural weathering[14].

3.5 Surface Analysis

Surface analysis of the substrate and sealant surfaces was performed to characterise the surfaces in terms of their chemistry, morphology and energy. The most useful technique proved to be contact angle analysis leading to a determination of surface energy values. These values provided a basis for predicting the potential adhesion and durability of resealed joints. Static contact angles were measured using four test liquids: water, formamide, ethylene glycol and DMSO. Surface energy measurements were then calculated using equation (8) derived earlier.

4 RESULTS AND DISCUSSION

4.1 Surface Energy Values

All of the surface energy values calculated are shown in Table 4.

4.2 Mechanical Test Data

The mechanical testing of joints made with contaminated substrates produced a complex series of results, especially in relation to the 'hybrid' nature of the test specimens which contained many materials and interfaces. A novel approach was developed in order to succinctly express the test data, especially for the large number of potential failure modes. Summary graphs of the general trends regarding key material properties such as failure mode, are given and correlated with the trends predicted from surface analysis in figures 6 and 7. A fuller account of the test data generated is provided elsewhere[15].

4.3 Effect of Surface Condition on Joint Performance
From an average of the surface energy values calculated for the contaminant sealant layers on all three substrates (thin layer contamination), the following order of adhesion performance and bond durability would be predicted :

highest surface energy   lowest surface energy

acr > 1-pu > 2-ps > BR > Bitumen > GP mastic > sil b

35.5 > 34.4 > 33.0 > 28.4 > 27.9   > 27.9   > 22.6   (mJm^{-2})

best predicted surface   worst predicted surface
for adhesion   for adhesion

If the mechanical testing results are examined, the following trend can be arrived at based on the failure modes and extensions obtained, for the actual order of adhesion on each contaminant surface:

2-ps > 1-pu > acr > BR > sil b > Bitumen > GP mastic
best surface for   worst surface for
adhesion   adhesion

It can be seen in general that the predicted order and the actual observed order of performance, while not identical, are quite similar.   Hence the surface energy characteristic of a contaminated surface is an important indicator of potential adhesion. However it does not explain the entire picture of potential adhesion which is why the above two orders of performance are not identical. For example, the effect of different sealant moduli is not accounted for and the influence of primer materials is also ignored.

Evidence has emerged from the surface analysis conducted on the thin-layer contaminated substrates that these surfaces exhibit both sealant and substrate characteristics.   This can be deduced from surface energy studies of these surfaces which indicate energy values somewhere in between the respective surface energy values measured for sealant and substrate.   An example of this is:

Anodized aluminium (clean)   37.2 mJm^{-2} surface energy

Sil b sealant   14.5 mJm^{-2} surface energy

Anodized aluminium/sil b   25.2 mJm^{-2} surface energy

However, chemical analysis of this surface using X-Ray Photoelectron Spectroscopy indicated that the surface chemistry was purely silicone, ie sealant. In the above trend it is clear from both the predicted and actual order of performance that traditional 'non-curing' oil-/rubber-based sealant contaminants provide the worst surfaces for adhesion. The surface energies are low and various oils and plasticisers used in their manufacture would also hinder the development of good adhesion.

It was noticed during preparation of the contaminated surfaces that when the polysulphide(2-ps) contaminant was cut back, the surface that remained was slightly rougher than that observed for the other sealants. This could be related to tear/cutting resistance and could be part of the explanation of why the polysulphide, as a contaminant, performs better than predicted from surface energy values alone.

4.4 Effect of the Thickness of Contamination on Joint Durability

A general comparison of the effect of the thickness of contaminant leads to the following observations:

• Thick-layer contaminants, unprimed.
General large-scale reduction in extensions at peak load for all sealants, associated with adhesive-type modes of failure. Thick-layer contamination should be treated with the upmost caution, especially non-curing contaminants such as GP mastic.
• Thick-layer contaminants, primed.
No significant improvement compared with unprimed thick layer. Again, rather poor sealant joint performance.
• Thin-layer contamination, primed.
Performance of sealants is reduced compared to clean substrates. However failure modes and extension generally remain acceptable (eg. extensions at peak load typically ≥ 100%). The long-term significance of sealing to this type of surface is unknown for joint performance. The non-curing contaminants provide the greatest problem, especially on porous substrates such as concrete.

Failure between reseal primer and resealing sealant occurred more often than expected. This may be because the presence of a sealant contamination detrimentally affects primers designed for clean substrates. This might suggest a need for primers designed specifically for contaminated surfaces.

The importance of the parameter 'extension at peak load' emerged from this work. It appears to be one of the most effective parameters to judge the adhesion and durability of sealants on different substrates. A significant drop in extension exhibited by a sealant on a given surface is an indicator of adhesion loss and reduced joint durability with that surface.

Table 4 Summary of Surface Energy Information on Substrates and Sealants

surface condition	Substrates anodised			powder-coated			concrete			cured sealant surface			un-cured sealant surface		
	γ^p	γ^D	γ^s	γ^p	γ^D	γ^s	γ^p	γ^D	γ^s	γ^p	γ^D	γ^s	γ^p	γ^D	γ^s
Clean Substrate Surface	13.0	24.2	37.2	19.9	16.5	36.4	unobtainable								
1 PU(R/C)	22.1	11.5	33.6	22.6	13.1	35.7	19.1	16.6	35.7	9.4	17.2	26.6	16.2	18.4	34.6
2 PU(R)													16.3	24.9	41.2
1 PS(R)													8.3	20.6	28.9
2 PS(R/C)	13.6	20.3	33.9	14.1	19.0	33.1	6.2	25.7	31.9	4.4	20.0	24.4	6.4	26.0	32.4
Sil o (R)													6.3	6.4	12.7
Sil b (R/C)	12.3	12.9	25.2	13.1	13.3	26.4	8.3	9.4	17.7	2.8	11.7	14.5	5.8	11.5	17.3
Acr (R/C)	20.3	15.2	35.5	17.8	16.4	33.2	23.0	14.2	37.2	16.0	12.2	28.2	19.6	15.4	35.0
BR (C)	15.4	13.4	28.8	14.1	16.2	30.3	5.8	19.7	25.5	2.6	22.0	24.6			
GP Mastic (C)	10.1	18.8	28.9	12.0	20.0	32.0	8.2	17.6	25.8	2.1	22.5	24.6			
Bitumen (C)							0.3	33.8	34.1	1.0	24.3	25.3			

γ^p = Polar component of surface energy; γ^D = Dispersive component of surface energy; γ^s = Total surface energy (mJm^{-2}) C = Contaminated surface; R = Reseal material

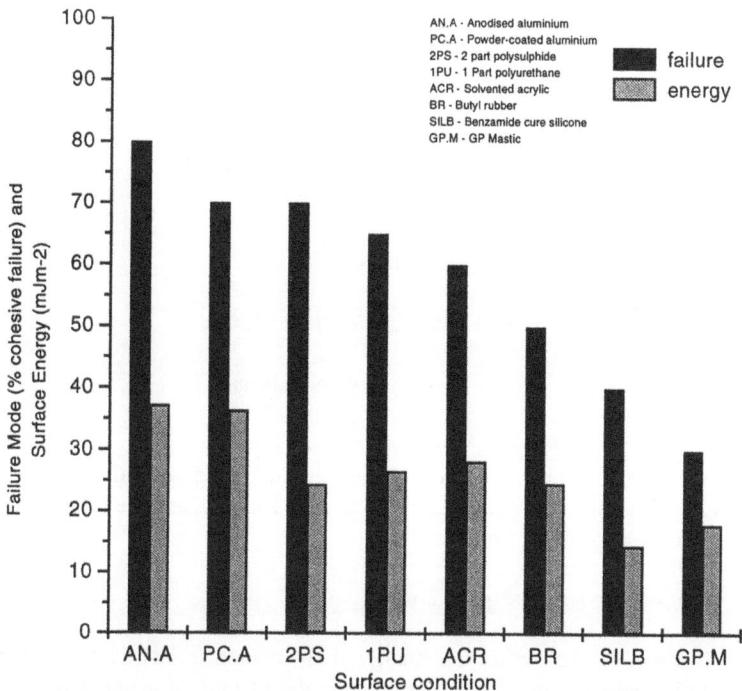

Fig.6. Influence of the energy of a surface on its adhesion performance

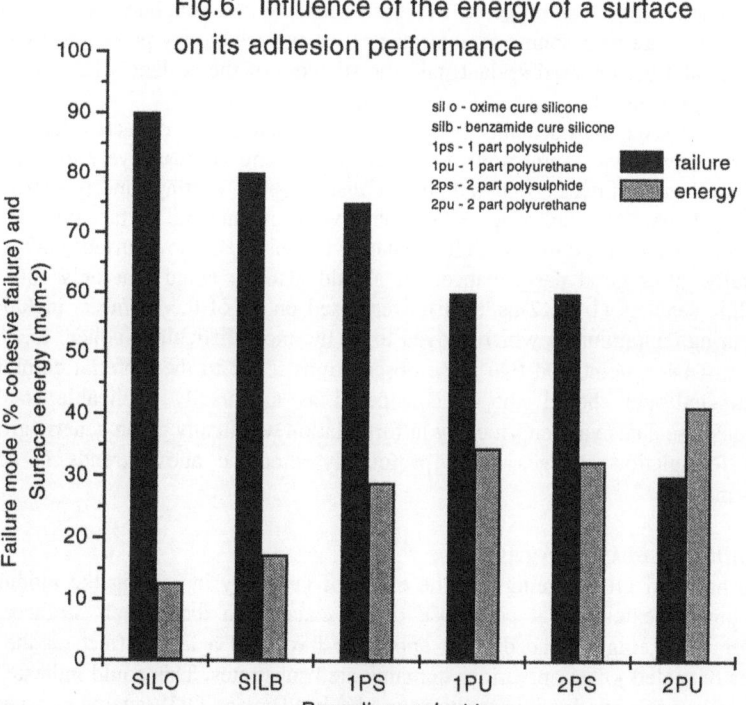

Fig. 7. Surface Energy/failure mode relationship for uncured 'resealing' sealants

4.5 Comparison of Performance of Resealing Sealants

From the surface free energy values calculated for uncured sealant surfaces the following order would be predicted for the best adhesion:

| lowest surface | greatest surface |
| energy | energy |

sil o < sil b < 1-ps <   2-ps < 1-pu < 2-pu

12.0 <   17.3   < 28.9 < 32.4   < 34.6 < 41.2   (mJm^{-2})

| best adhesion | worst adhesion |
| potential | predicted |

If the mechanical testing results are analyzed the following generalised adhesion and durability performance order can be arrived at from summarising the failure modes and extension at peak load performance:

sil o > sil b > 1-ps; 1-pu > 2-ps > 2-pu

Obviously the lower the surface free energy of an uncured sealant the better its wetting characteristics will be and hence the greater the chance of good adhesion. This generalisation must be tempered by: the chemistry and roughness of the substrate surface; the relative amounts of polar/dispersive surface energy possessed by both the substrate and the uncured sealant; and the stiffness of the sealant. These factors will have a significant effect on a sealant's adhesion potential.

Conversely the 2-ps sealant gave worse performance than expected because of its relatively low polar component of surface energy and comparatively high modulus. The 2-pu has the greatest uncured surface energy (poor wetting) and is a stiff sealant (high modulus). Both silicone sealants have very low uncured surface energies (good wetting). The 1-ps is a low modulus sealant with reasonably low surface energy which generally gives good performance. It should also be noted that only four of the resealing sealants (1-pu, 2-ps, sil b) were used on all of the surfaces, including the non-curing contaminants which proved to be the most difficult to adhere to.

It must be remembered that these observations relate to the material combinations investigated and should not be considered as universally applicable statements. Obviously sealants vary enormously in formulation within any given generic group and these formulation changes could profoundly alter the above trends in adhesion performance.

4.6 Influence of QUV Ageing

2,000 hours of QUV ageing had the effect of generally increasing the modulus and reducing the extension at peak load of all sealants on the various surfaces. This accelerated ageing regime did not appear to have had a large effect on the failure modes of sealed joints on various contaminated substrates. This could indicate that 12 weeks weathering under the conditions used is insufficient. QUV ageing in general may be inadequate to fully test the potential of a sealed joint, at least in the absence of imposed mechanical strain.

This accelerated weathering regime used was disappointing in its contribution towards understanding sealant adhesion and durability. The main reason for this was probably due to the lack of any imposed mechanical strain on the test joints during the accelerated ageing, since other researchers[16] have shown that cyclic mechanical stress is a key element in an accelerated ageing regime. These sets of results tend to reinforce that observation. Further work on assessing long term adhesion of sealants must include a mechanical stress component. The selection of an appropriate strain/stress level and cycling rate to use require careful selection, especially if real building movement are to be simulated.

5 Summary Observations

This study demonstrated that scientific surface analysis could be used as a fairly accurate predictor of the adhesion and potential bond durability of sealants on different substrate surfaces.

The effect that different sealant contaminants have on the adhesion of resealing sealants can be anticipated, with reasonable accuracy, from surface energy measurements. High energy surfaces with a large proportion of polar surface energy are the optimum surface conditions to promote initial adhesion. It was also found that better adhesion was gained to curing contaminants than to non-curing contaminants when resealed by a curing contaminant. The presence of sealant residues inevitably leads to a reduction in joint extension and bond performance, in proportion to the amount of residue present and its surface energy.

Resealing sealants which possess low surface free energies in the uncured state have a greater chance of effectively wetting out a surface. The best bond durability was found with joints prepared with high surface energy substrates and low uncured surface free energy sealants.

Acknowledgements

The authors express their appreciation to the UK DOE/EPSRC LINK CMR Management Committee and all of the sponsors and partners for their support of the project.

References

1. Hutchinson, A.R. (ed). Resealing of Buildings : A Guide to Good Practice; Butterworth-Heinemann, 1994.
2. Wolf, A. Studies of the Ageing Behaviour of Gun Grade Building Joint Sealants; Polymer Degradation and Stability,23, 1989, pp.139-163.
3. Wake, W.C. Adhesion and formulation of Adhesives; Applied Science Publishers Ltd., 1976, pp.21-23.
4. Allen, K.W. Adhesion and Adhesives — Some Fundamentals; Adhesives and Consolidants Conference Proceedings, 1984, p.5.
5. Allen, K.W. Some Reflections on Contemporary Views of Theories of Adhesion; Int.J. Adhesion and Adhesives, April 1993, p.69.

6. Dalal, E.N. American Chemical Society, 3, 1987.

7. Wu, S. Polymer Interface and Adhesion; Marcel Dekker, 1982.

8. Good, R.J. and Girifalco, L.A. J.Phys.Chem., 64, 1960.

9. Kaelble, D.H. Physical Chemistry of Adhesion; Wiley-Interscience, New York, 1971.

10. Kinloch, A.J. J.Material Science, 17, 1982, p.617.

11. Kinloch, A.J. Durability of Structural Adhesives; Applied Science Publishers, 1983, p.9.

12. Marechal, J.C. and Ghaleb M., Duarbilite des assemblages colles par mastic silicone - application aux vitrages exterieurs, CSTB, No. 297 Mars 1989.

13. Allen K.W., Hutchinson A.R., Pagliuca A., Curing Characteristics of Construction Sealants, Int.J. Adhesion and Adhesives, April 1994.

14. Fedor, G. and Brennan, P. Correlation of Accelerated and Natural Weathering; Adhesives Age, May 1990.

15. Pagliuca, A. and Hutchinson A.R., Adhesion Properties of Sealants in Resealed Joints, Science and Technology of Building Seals, Sealants, Glazing and Waterproofing: 6th Volume ASTM STP 1286, J.C. Myers, Ed., American Society for Testing and Materials, Ft. Lauderdale, 1996.

16. Karpati, K.K. Laboratory Fatigue Test of a Two-part Polysulphide Sealant Correlated to Outdoor Performance; Durability of building Materials, 1987.

Appendix 1 **Preparation of Surfaces**

The contaminated surfaces were prepared as follows.

(1) All substrates (except powder-coated aluminium) were cleaned using a paper cloth and 1,1,1 trichloroethane. The powder-coated aluminium was cleaned using white spirit instead of Genklene, in common with normal practice. This is because Genklene will damage a powder-coated surface if allowed to stay on the surface for a few seconds.

(2) The substrates were then primed according to the relevant sealant manufacturers instructions.

(3) A layer of contaminant sealant was applied using a specially designed comb to leave a sealant layer approximately 4mm thick on the substrate.

(4) A ten-week curing/ageing regime was then used to age the contaminant. This ten-week regime consisted of:

   (i)   six weeks outside curing

   (ii) four weeks curing in an oven at 60°C.

(5) After the ten-week curing/ageing period the contaminated substrates were ready to be cut back and cleaned.

Two different cutting back techniques were adopted in order to generate the two levels of sealant contamination.

(6) (a) A thin layer (<0.1mm) of sealant residue was produced in order to simulate an effective cleaning operation in the field. This involved cutting back the sealant contaminant with a chisel as close as possible to the substrate surface (but avoiding damage to the substrate surface itself). The substrates were then gently cloth-wiped three times with Genklene (white spirit for powder-coated aluminium) and left for two days at ambient temperature and humidity before being used for joint

preparation.

(b) A thick layer (0.5mm) of sealant residue was produced, again by cutting back the sealant contaminant with a chisel. However, the depth of cut was controlled by placing the contaminated substrate in a purpose-designed jig which resulted in an accurate 0.5mm layer of sealant remaining on the substrate. These thick layer contaminated substrates were not solvent-cleaned but were left at ambient temperature and humidity for two days prior to being used for joint preparation.

9 THE CONTRIBUTION OF RESEARCH TO STANDARDISATION

J.C. BEECH
Building Research Establishment,
Garston, Watford, UK

ABSTRACT

In the past standardisation for building and construction products has often been based on experience of the relevant polymer technology, with the major objectives the achievement of adequate material quality and quality control in the manufacturing process.

International standardisation is now firmly based on performance specifications, and test methods are now being developed to support this approach. Perhaps the most difficult task is to predict durability from the results of laboratory tests.

The work of ISO Technical Committee TC59/SC8 has reached the stage of considering durability assessment, and a Working Group (WG6) has been set up to undertake the task of developing a suitable test method.

This paper reviews the contribution that research has made, and continues to make, to the development of performance-related test methods for assessing the potential performance of building sealants, with particular reference to their movement capability and durability. It draws on experience of research in the United Kingdom and Europe which supports standardisation and the harmonisation of standards, and considers the prospects for inclusion of a "durability" test within the corpus of test methods which has been developed by ISO and adopted by CEN as part of the European standardisation of building and construction products.

Keywords: durability, movement accommodation, sealants, standards

INTRODUCTION

In the past, specifications for sealants have often been developed for specific generic classes of products, and the test methods stipulated for their assessment have been based upon the well-developed technology of the polymers from which the sealants were manufactured.

The British Standards for polysulphide and silicone based sealants, in their earlier editions [1,2,3], are typical of these. All contain relatively simple tests to check the ability of the relevant products to maintain cohesion and adhesion to a substrate after subjecting specimens to maintained extension (and in one case to compression also). The essential test conditions are different for the three classes of sealant: viz the amounts (%) and rates of extension, and the durations of maintained extension. This has been discussed in detail in an earlier paper, interested readers are referred to reference [4].

The point to be noted here is that all these products are intended to fulfil essentially the same functions in the same range of service conditions.

When work to standardise sealants for building and construction started in ISO Technical sub-committee TC59/SC8 in the mid-1970s, a broad strategy was soon established: to work towards performance-based specifications which would draw upon methods of test related to the functions of the products and to the service conditions in which these functions must be performed.

In the years since, steady progress has been made towards this objective: a corpus of test methods has been created, embodied within a Classification document [5], which contains the elements of two performance-based specifications, for glazing sealants and for sealants used to seal movement joints in buildings.

This major international collaborative effort has been underpinned by research of two kinds. First, by individual research projects in universities, research organisations and in commercial companies to establish the bases for test methods; for example, the influence of specific environmental agents on the long-term performance of various products.

Secondly, once a proposal for a test procedure has been elaborated, "Round Robin" test procedures have been conducted to check the feasibility of the method and to validate, as far as possible in the short term, the conditions to be stipulated in the test method.

This paper does not attempt a comprehensive review of the research in the field of sealants standardisation; rather it discusses some examples within this activity in which research has played an essential role, within the writer's experience of standardisation in the ISO and BSI sealants standards committees.

RESEARCH AND STANDARDISATION

The examples chosen for consideration relate to three important aspects of the performance of sealants used in building and construction: Movement Capability, the ability of the product to withstand repeated joint movements through its service life; Water Resistance; and the assessment of Durability, perhaps the most demanding challenge facing the materials scientist researching the performance of building materials, but imposing some problems peculiar to joint sealants.

MOVEMENT CAPABILITY

Correct joint design requires that joint widths should be calculated from a knowledge of the movement capability, which must therefore be regarded as a key performance property.

Test methods to assess this property usually include cycles of extension and compression of samples, which may be performed with simultaneous variation in temperature. They aim thereby to simulate, however crudely or approximately, the conditions to which the sealant is subjected in a movement joint: as the temperature falls the sealant is placed under increased extension, and is eventually compressed with rise in temperature.

Early versions of this kind of test are found in sealants specifications issued in the USA in the 1960s[6], deriving from the research of Hockman at the National Bureau of Standards. Strangely, in view of his influence on the development of sealants test methods, Hockman appears never to have published the research results upon which he based the proposals for this classic test.

The current version of the Hockman test is found in ASTM standards[7]. It should be noted that, as when first published, this test does not conform precisely to the simulation of actual joint movement and temperature variation described above. The essential features of the method are: immersion in water followed by heating of the compressed specimens; 10 cycles of extension and compression at standard conditions of temperature and relative humidity; and finally, nine cycles of compression at 70ºC and extension at -26.1ºC. While the compression is maintained for 16-20 hours, the low temperature phase is limited to the time needed to extend the specimens by the required amount, after which they are allowed to warm up to standard temperature with the extension maintained.

When the sealants committee of the British Standards Institution considered the requirements for a movement capability test in the late 1970s, account was taken of the results of tests at BRE and elsewhere in which specimens of a range of commercial sealants were exposed to weather on joint movement simulators. These apparati were actuated to impose cyclic movement on the sealants specimens by the expansion and contraction of rigid PVC bars exposed to solar radiation and to variations in ambient air temperature. As a result, the rates of movement of the sample "joints" were very rapid by comparison with the range of rates measured on actual building joints.

The effect of this was to produce relatively rapid failure in specimens of those products which were predominantly plastic or plasto-elastic in properties; while sealants which were essentially elastic in character performed well, as they would be expected to do in building joints which experienced rapid thermal movements.

A method of test was devised,therefore, which enabled sealants to be classified according to the amount of joint movement they would withstand, but the rate of movement in the test also depended on this movement amplitude [8]. There were six levels of movement capability, which ranged from M1 (capable of tolerating total movement of 10% of the joint width, with the rate of movement in the test being 4 mm/h) to class M6 (for which the movement capability was 60%, and samples were extended and compressed in the test at 24 mm/h). For all these classes, the test prescribed compression at 70ºC and maintained extension at -15ºC.

A test programme was carried out at BRE to evaluate this method of test, which as a "Draft for Development" was required to be assessed by experience in use over a period of five years before a decision whether to adopt it as a British Standard could be taken.

Despite the complexity of the "DD69" test it was found that while it successfully discriminated between high-performance elastomeric products and those of the lowest movement capability, it was less successful in identifying differences in performance among the elastomeric products. Thus elastomeric products including one-part polysulphides, polyurethanes and low-modulus silicones all achieved M5 or M6 classification, none of the sealants tested falling in the intermediate classes M2, M3 and M4.

Although the BRE assessment was based on a small number of products it did call into question the theoretical basis for the methodology of the test. A more serious practical obstacle to its adoption as a British Standard was the expense of the equipment required to perform the test, which was required to have the capability to operate at six different rates of movement. Clearly no commercial test laboratory was likely to invest in this equipment without some confidence that the test would eventually be adopted as a standard test for sealants.

Before the five year period of review for DD69 had elapsed, work had started by ISO/TC59/SC8 Technical Sub-committee on "Jointing Products" to tackle the development of a new movement capability test. At this time both the UK and the USA were participating members, though neither pressed for their own test methods to be adopted. In the case of the ASTM test, there was then at least 20 years experience of its successful use in the USA and elsewhere. However, the position of the USA was significantly different from that of the majority of European members, in view of the likely acceptance by CEN of the standards prepared by ISO for sealants, which would then become mandatory both for members of the European Community and EFTA.

The initiative for producing a proposal for a test method for consideration by the technical committee was taken by the representatives of the French and German member bodies, and after discussion regarding the details of the test procedure, a draft was eventually agreed on which the testing of typical products was to be based.

A limited test programme was carried out by a few members of the committee,which did not meet the criteria for "Round Robin" tests stipulated by ISO. Nevertheless, evidence was obtained that the test gave results which broadly reflected the known performance of the sealants tested in service; the technical committee agreed to the adoption of the method as an ISO test[9].

Wolf[10] made a comparison of the results of this test with the performance of a range of commercial sealants after 6-7 years of exposure to weather under mechanical strain. His research addressed a point of contention during the discussions in the ISO committee which led to the issue of the standard: whether there was a need to include compression of the specimens at 70ºC as part of the test procedure. the correlations presented by Wolf indicated, in his view, that inclusion of compression helped to identify those sealants which failed during outdoor weathering owing to high compression set. Wolf's work has made a significant contribution to the development of tests for durability, which we consider next.

It should be noted that work continues in the ISO committee to attempt to extend the method of ISO 9047 to the assessment of sealants of movement capability greater than 25%. Initial work at BRE to examine this possibility (unpublished) has indicated that while the test may be valid for products of 35%, it produces anomalous results when used to test sealants with a purported movement capability of 50% or more.

An account of the study of the movement capability of sealants must include reference to the work of Klara Karpati, whose work at the National Research Council, Ottawa, made a notable contribution to our understanding of sealant performance. Her many published papers are too numerous to list here, and despite the originality of her work it does not appear to have influenced the writing of sealants standards directly. Her demonstration that a simple vice could be used adequately to simulate the effects of exposure of specimens on a complex weathering/movement apparatus is likely to prove of value in durability studies now in progress.

DURABILITY

A comprehensive review of the many studies of the effects of natural and artificial weather on the properties of building sealants, and of the attempts to establish correlations between the two kinds of data, would clearly be beyond the scope of this paper. In addition to the previously cited paper[10], Wolf published in 1988 a valuable "State of the Art" review [11] of work on sealant durability, and since that date a number of relevant papers have been presented at ASTM symposia on sealants and related products, and elsewhere.

In 1991 a RILEM technical committee was set up specifically to study this topic, under whose auspices this seminar is of course taking place. The work of RILEM TC 139-DBS complements that of the ISO technical committee to which previous reference has been made, and which has itself formed a working group to make recommendations regarding an ISO test method for durability.

Although weathering machines utilising xenon-arc lamps to simulate the effects of solar radiation have become well-established in many areas of materials testing, the costs of such apparatus, and the expense of maintenance, have militated against their use in sealants research, and encouraged the use of cheaper alternatives, such as machines based upon the use of fluorescent tubes. Studies to determine the effects of different weathering apparatus on surface changes of sealants, such as Fedor[12], have tended to produce similar results with xenon-arc and fluorescent tube based equipment.

With regard to the latter types of artificial weather machines, debate has focused on preference for using fluorescent tubes emitting ultra-violet radiation in the "A" or "B" bands. The former emit radiation closer to the UV spectrum of the sun at the surface of the Earth, with peak radiation at a wavelength at 340 nm. UV(B) lamps have a spectrum which corresponds to the narrow band of solar radiation which is sufficiently energetic to cause bond sission in many polymeric materials when these are exposed long term to the sun. These lamps have peak radiation intensity at 313 nm.

While there is a risk that use of UV(B)-based artificial weathering will induce reactions unlikely to occur under real service conditions, it is possible to produce degradation broadly typical of what occurs in sealants exposed to natural weathering much more rapidly than with equipment using UV(A) lamps.

Accordingly, extensive studies carried out at BRE in the UK in recent years have employed QUV apparatus[13] with UV(B) lamps and a test chamber temperature of 70ºC, alternating with condensation phases at 50ºC[14,15,16]. Attempts to correlate the effects of such laboratory exposures with data from unstressed specimens exposed to natural weather have been based on measurements of tensile properties including modulus and extensibility.

The results of these attempts at correlation, using data from polysulphide, silicone and polyurethane-based products, and drawing upon sealant specimens exposed for up to 6.5 years in temperature, tropical and middle-eastern climates, have been promising. However, they have been obtained mainly with measurements of only one property, namely modulus. Validation of the use of this type of weathering machine as part of a standard durability assessment clearly requires further confirmation that significant correlations can be demonstrated between changes in other properties of specimens, when compared with the results of actual weather exposure.

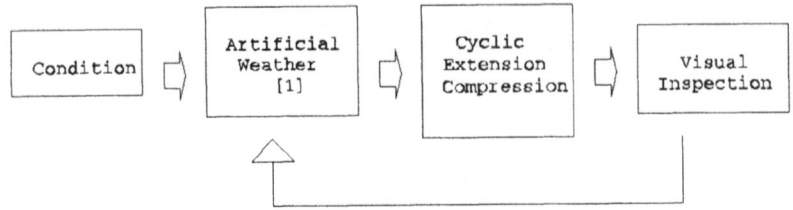

Repeat this cycle "N" times

[1] Artificial weather; 28 days of UV/water exposure

[2] Cyclic Extension/compression to ISO 9047

Figure 1

In addition, it will be desirable to determine performance in cyclic movement tests which simulate joint movement, in conjunction with simulated weather exposure. This is the basis of a proposal put forward by the Working Group 6 of the ISO committee[17], and currently under discussion. It is intended to use the test method of ISO 9047 as the "cyclic movement" element of this evaluation, for elastomeric products have movement capabilities of up to 25%. (An alternative ISO test is available for the movement capability of sealants with predominantly plastic properties[18]. This would appear to be an appropriate cyclic movement element for a durability test to assess such products).

The particular merit of the proposed "durability" test is that it can incorporate whichever cyclic movement element is appropriate to the properties and function of the sealant. As far as the weather simulation element is concerned, provision is made in the draft for a choice of different artificial weather machines to be made, and the precise details of the solar radiation/water simulation may also be varied to suit the requirements or availability of test equipment.

The test procedure is shown in diagrammatic form in Figure 1. As will have been inferred from the foregoing description, the test method has been drafted in a form which will enable collaborative evaluation to be carried out, in which the effects of varying the details of the experimental procedure may be assessed. Thus, although it is envisaged that many of the international researchers participating in this exercise will choose to employ the QUV apparatus, with radiation from UV(B) lamps, and test exposure according (eg) to the standard cycle of ISO 4892-3 [19], other laboratory simulations of natural weather will not be arbitrarily excluded from consideration.

The ISO draft for a "durability" assessment test has been available to national standards bodies for over 12 months, and it is not yet clear how many test laboratories within the countries active in this standardisation effort have commenced experimental work to evaluate this test, or are likely to be involved. Clearly, this is an area of standards work in which collaborative research will be essential to future progress towards an agreed test method.

It may be appropriate also to mention briefly a number of ongoing research projects which are contributing to the data base upon which an agreed method of test for durability assessments of sealants may eventually be founded and validated. Two of these are taking place under the auspices of the RILEM Technical Committee 139-DBS; it is hoped that a review of progress on both these studies will be possible at this seminar.

The first is taking place in the USA at the South Florida Test Service, under the direction of Lesley Crewdson and with the collaboration of other RILEM Technical Committee members. It aims to investigate the correlation between sealant properties after periods of exposure to weather at various sites around the world, and similar data obtained after laboratory weathering.

The second study has similar aims, but is concerned particularly with adhesion aspects of the performance of silicone-based sealants. This is being carried out by Dr W Gutowski and his colleagues at the Commonwealth Scientific and Industrial Research Organisation (CSIRO) in Highett, Victoria, with financial support from the Australian sealants industry.

This study also seeks to investigate the influence of a factor which has been identified by several researchers, including Wolf[10] and Sandberg[20], as of importance in the performance of sealants in joints: namely, the state of stress of the material during exposure to weather. Accordingly, it is intended to expose sealants specimens under constant tensile stress at various levels to weather in different geographical locations.

A laboratory test to assess potential long term performance by simulation of the conditions of service is usually constrained by practical limitations, and is therefore in most respects a simplification of the total environment in which a sealant must fulfil its functions. For example, the ISO draft method described above subjects test specimens <u>sequentially</u> to simulated weather factors (Solar radiation, temperature variations, water), and to simulated joint movement. An important aspect of movement, the effect of maintained tensile stress, may be imposed upon samples simultaneously with these weather factors. The ultimate test procedure for durability assessment is to embrace all the relevant factors which are known to affect performance and durability within a single dynamic test.

This would appear to require the development of an elaborate test apparatus in which suitable sealant specimens may be simultaneously subjected to simulated stressing due to joint movement and to simulation of weather, including any seasonal variations and differences related to the specific climate which are thought to influence long-term performance in service. Considerations of the cost of the equipment, and of the extreme difficulty of validating the results of such a test, have deterred all but the most ambitious from attempting to pursue this objective. Nevertheless the advantages of such an all-embracing test method, which would attempt to take account of synergistic effects of movements and weather, is apparent.

A European collaborative project led by Dr Shaun Hurley of Taywood Engineering, UK, is in progress with the objective of developing a comprehensive test method for the durability of building joint sealants. Details were given of this project at an early stage at a BRE seminar in 1992, and an update is promised in the paper Dr Hurley is presenting at the present seminar[21].

While the objective is extremely ambitious, and the work of validating the results of the test which is ultimately proposed will be arduous, it is to be hoped that the results of the various active research projects which are complementary to it will contribute to the successful completion of the work.

WATER RESISTANCE

The ISO classification document ISO 11600 makes reference to two ISO test methods which relate to the influence of water on the performance of glazing and construction sealants. The test relevant to elastomeric products used in building joints provides for the measurement of adhesion/cohesion properties with maintained extension after 4 days immersion in water[22]. This is an adequate assessment of the effect of intermittent and short duration exposure to water which may be expected to occur in joints in the external walls of buildings. However it is totally inadequate for the evaluation of sealants which must withstand longer periods of exposure to water.

Two further degrees of water exposure may be identified in this regard: for joints in horizontal surfaces subject to flooding or ponding; and for joints subject to continuous immersion such as in water-retaining structures.

It was accepted by the ISO committee that further development was required of a test method which would take account of these service requirements. An ad hoc group was set up to consider the available experimental data pertaining to long-term water resistance. Considerable research effort had been expended in the USA, where ASTM Committee C-24 had set up a task group of Technical Sub-committee C24.32.00 to deal with prolonged water exposure of chemically-curing sealants.

In the UK a project was active, managed by the Construction Industry Research and Information Association (CIRIA), with participation by the Water and Sealants Industries, to study the condition of sealants in reservoirs and other industrial aqueous media, and to make recommendations for a test method to form the basis for an approvals system for sealants used in water-retaining structures.

In connection with this work a test programme was undertaken at BRE to study the effects of prolonged immersion on the tensile properties of sealants. The results of this research[23] formed the basis of the specification for sealants used by the UK Water Industry which was later published by the then Water Services Association[24].

The method of test required the measurement of tensile properties of specimens which had been immersed in water at ambient temperature for up to 12 months. The duration of the test posed obvious problems for users and suppliers of sealants needing to obtain approval for the use of products in the water industry. Attention was therefore focused on finding a means of reducing the duration of the test procedure, which had been tackled in the ASTM studies[25] by increasing the temperature of the immersing water from ambient to 50ºC, on the assumption that an Aarhenius type relationship existed between the rate of change or physical properties (primarily the adhesion of the sealant to the substrate and the softening of the sealant with water adsorption) at the two temperatures.

The ISO ad hoc group considered the available experimental data, but was unable to agree on a proposal for a test method in the absence of a demonstrated correlation between the sealant properties after immersion in water at the ambient and elevated temperatures.

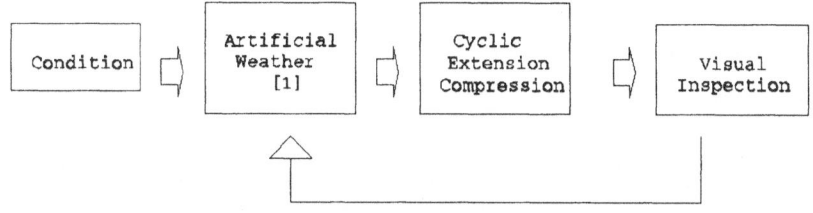

Repeat this cycle "N" times

[1] Water Immersion for 21 days at 23°C, or 40°C, or 50°C

[2] Cyclic Extension/compression at 50% of the amplitude for

movement capability determination

Figure 2

The ISO committee later set up a working group (WG7), with the present writer as Convenor, to take this work further. BRE had undertaken a further test programme to study the changes in tensile properties which occurred after immersion in water at 50ºC. The results showed clearly that there was no correlation between key performance indices such as extensibility and modulus, after immersion at the respective temperatures.

Discussion in ISO WG7 elicited the observation that a marked change in the adhesion characteristics of some sealants appears to occur with water immersion as the water temperature is raised from 40ºC to 50ºC. It is tentatively concluded, therefore, that an increase in the water temperature to 50ºC may be invalid, and that acceleration of the changes in sealants due to water immersion during this test may need to be limited to what can be achieved by raising the water temperature to 40ºC.

Taking account of the experimental data now available, a method of test has been drafted for consideration by the ISO WG7 and the main committee. This is based, by analogy with the "durability" test described above, on sequential exposure of sealant specimens to water immersion following by a small number of cyclic movements, the amplitude of which is related to the stated movement capability of the product.

The test procedure is shown diagrammatically in Figure 2. It has some features in common with the ASTM C24 draft test method, and also makes provision for the water immersion phases to be carried out at a number of different temperatures. The intension is to encourage testing by participants in the standards making activity, using a wide range of elastomeric sealants intended for use in aqueous environments, and using the three stipulated temperatures for water immersion. As well as allowing an assessment of the method from the performance of a range of products of known performance in service, it is hoped that the collaborative test programme will also indicate if there is correlation between performance at the different water temperatures stipulated in the test method.

As with the draft test method for "durability", this test procedure has been drafted in a form which will promote further investigation of the requirements for a predictive test. In particular there is a need for more experimental data to establish the relationship between the number of water resistance "cells" which sealants will withstand in the test and their ability to perform satisfactorily in wet service environments of different degrees of severity. (It should be noted that service environments may include, in addition to structures storing potable water, joints exposed to sewage and in marine environments. The test method must be suitable for adaptation for use in evaluating sealants intended for such applications).

achieved for all performance requirements of building sealants, it is perhaps in the field of validation of the available test methods that the major challenge to researchers remains.

CONCLUSION

The foregoing account of some aspects of standards work for construction sealants over the past 20 years, though limited to the writer's experience, illustrates well the growing contribution which research has made in this field. It excludes, of course, the extensive work of evaluation done within the companies of the sealants industry, and although this information is not generally in the public domain, representatives of the Industry have played an active and vital role in developing national and international standards during the period reviewed.

Although it will be clear from the description of ongoing work that much remains to be done before agreed international standards are The significant growth in interest in the technology of sealants in recent years is encouraging, and gives confidence that the challenge will be faced successfully.

REFERENCES

1. Specification for Two-part Polysulphide-based Sealing Compounds for the Building Industry. BS 4254 : 1967. British Standards Institution.

2. Specification for One-part gun-grade polysulphide-based sealants. BS 5215 : 1975. British Standards Institution.

3. Specification for Silicone based building sealants. BS 5889 : 1980. British Standards Institution.

4. Beech, J C. : Test methods for the movement capability of building sealants. The "state of the art". Materials and Construction, 18, (108), 473-482, (1985).

5. International Standard ISO11600 Building construction-Sealants-Classification and requirements. International Standards Organisation (1993).

6. American Standard Specification for Polysulfide-base Compounds for the Building Trade. A116.1 - 1960. American Standards Association Inc. (1960).

7. Standard Test Method for Adhesion and Cohesion of Elastomeric Joint Sealants under Cyclic Movement. ASTM C719-79. American Society for Testing and Materials, Philadelphia, (1979).

8. Draft for Development : Method for classifying the movement capability of joint sealants. DD69 : 1980. British Standards Institution (1980).

9. International Standard ISO 9047 "Determination of Adhesion/cohesion Properties at Variable Temperature" International Standards Organisation. (1988).

10. Wolf, A T. : The Development of an International Standard on a Movement Capability Test Method for Building Joint Sealants. Kautschuk + Gummi-Kunststoffe, 43, Nr11/90, (November 1990).

11. Wolf, A T. : Studies of the Ageing Behaviour of Gun-grade Building Joint Sealants - The "State-of-the-Art", Polymer Degradation and Stability, 23, 135-163 (1989).

12. Fedor, G R. : Usefulness of Accelerated Test Methods for Sealant Weather. Science and Technology of Building Seals, Sealants, Glazing and Waterproofing, ASTM STP 1200, Jerome M Klosowski, Ed., American Society for Testing and Materials, Philadelphia, (1992).

13. Operating Manual for the QUV Accelerated Weathering Tester, The Q-Panel Company, 26200 First Street, Cleveland, Ohio, USA.

14. Beech, J C and Beasley, J L. : Evaluation of Cure and Durability of Building Sealants. Science and Technology of Building seals, Sealants, Glazing and Waterproofing, Second Volume, ASTM STP 1200, J M Klosowski, Ed., American Society for Testing and Materials, Philadelphia, (1992).

15. Beech, J C and Beasley, J L. : Further Studies of Cure and Durability of Building sealants. Science and Technology of Building Seals, Sealants, Glazing and Waterproofing, Third Volume, ASTM, STP 1254, J Myers, Ed., American Society for Testing and Materials, (1993).

16. Beech, J C and Beasley, J L. : Effects of Natural and Artificial Weathering on Building Sealants. Science and Technology of building Seals, Sealants, Glazing and Waterproofing, ASTM, STP 1243, D Nicastro, Ed., American Society for Testing and Materials, Philadelphia, (1994).

17. Draft prepared by J M Klosowski, Convenor of ISO/TC59/SC8/WG6 Proposed ISO Test Method for Durability of Sealants subject to Outdoor Weathering as simulated in an Artificial Weathering Machine. ISO Committee Document (1993).

18. International Standard ISO 9046. Determination of Adhesion/cohesion properties at constant temperature. International Standards Organisation (1987).

19. Draft International Standard ISO 4892-3 Methods of Exposure to Laboratory Light Sources. Part 3: Fluorescent Lamps. International Standards Organisation (1991).

20. Sandberg, L B. : Comparisons of Silicon and Urethane Sealant Durabilities. J Materials in Civil Engineering, 3, (4), 278-291, (1991).

21. Hurley, S A. : The Prediction of Long-term Sealant Performance from Dynamic Accelerated Weathering. Paper to be presented at BRE/RILEM Seminar on the Durability of Building Sealants, 11-12 October 1994.

22.   International Standard ISO 10590.   Building construction-Sealants-Determination of adhesion/cohesion properties after immersion in water. International Standards Organisation (1991).

23.   Beech, J C and Mansfield, C. : The Water Resistance of Sealants for Construction, <u>Building Sealants; Materials, Properties and Performance,</u> ASTM, <u>STP 1069</u>, Thomas F O'Connor, Ed., American Society for Testing and Materials, Philadelphia, (1990).

24.   Water Industry Specification No 4-60-01: Specification for Building and Construction Joint Sealants. Water Research Centre, Medmenham, (1991).

25.   ASTM New Standard Test Method for Durability of Sealants for Underwater or Wet Applications.   Draft No 3 (May 1992).
(NB: This is a private document of the ASTM C24 Committee)

Index

This index is based on keywords assigned to the individual chapter as its basis. The numbers are the page numbers of the first page of the relevant chapter.

RILEM PUBLICATIONS

Information on RILEM publications can be obtained from: E & F N Spon, 2-6 Boundary Row, London SE1 8HN, Tel: International + 171-865 0066, Fax: International + 171-522 9623; or Chapman & Hall Inc, 115 5th Avenue, New York, NY 10003, USA, Tel: (212) 244 6412, Fax: (212) 268 9964.

RILEM Reports

1   **Soiling and Cleaning of Building Facades**
    Report of Technical Committee 62-SCF. *Edited by L. G. W. Verhoef*
2   **Corrosion of Steel in Concrete**
    Report of Technical Committee 60-CSC. *Edited by P. Schiessl*
3   **Fracture Mechanics of Concrete Structures - From Theory to Applications**
    Report of Technical Committee 90-FMA. *Edited by L. Elfgren*
4   **Geomembranes - Identification and Performance Testing**
    Report of Technical Committee 103-MGH. *Edited by A. Rollin and J. M. Rigo*
5   **Fracture Mechanics Test Methods for Concrete**
    Report of Technical Committee 89-FMT. *Edited by S. P. Shah and A. Carpinteri*
6   **Recycling of Demolished Concrete and Masonry**
    Report of Technical Committee 37-DRC. *Edited by T. C. Hansen*
7   **Fly Ash in Concrete - Properties and Performance**
    Report of Technical Committee 67-FAB. *Edited by K. Wesche*
8   **Creep in Timber Structures**
    Report of TC 112-TSC. *Edited by P. Morlier.*
9   **Disaster Planning, Structural Assessment, Demolition and Recycling**
    Report of TC 121-DRG. *Edited by C. De Pauw and E. K. Lauritzen*
10 **Applications of Admixtures in Concrete**
    Report of TC 84-AAC. *Edited by A. M. Paillere*
11 **Interfaces in Cementitious Composites**
    Report of TC 108-ICC. *Edited by J.-C. Maso*
12 **Performance Criteria for Concrete Durability**
    Report of TC 116-PCD. *Edited by H. K. Hilsdorf and J. Kropp*
13 **Ice and Construction**
    Report of TC 118-IC. *Edited by L. Makkonen*

RILEM Proceedings

1   **Adhesion between Polymers and Concrete. ISAP 86**
    Aix-en-Provence, France, 1986. *Edited by H. R. Sasse*
2   **From Materials Science to Construction Materials Engineering**
    Versailles, France, 1987. *Edited by J. C. Maso*
3   **Durability of Geotextiles**
    St Rémy-lès-Chevreuses, France, 1986
4   **Demolition and Reuse of Concrete and Masonry**
    Tokyo, Japan, 1988. *Edited by Y. Kasai*
5   **Admixtures for Concrete - Improvement of Properties**
    Barcelona, Spain, 1990. *Edited by E. Vazquez*
6   **Analysis of Concrete Structures by Fracture Mechanics**
    Abisko, Sweden, 1989. *Edited by L. Elfgren and S. P. Shah*
7   **Vegetable Plants and their Fibres as Building Materials**
    Salvador, Bahia, Brazil, 1990. *Edited by H. S. Sobral*
8   **Mechanical Tests for Bituminous Mixes**
    Budapest, Hungary, 1990. *Edited by H. W. Fritz and E. Eustacchio*

RILEM Recommendations and Recommended Practice

RILEM Technical Recommendations for the Testing and Use of Construction Materials

Autoclaved Aerated Concrete - Properties, Testing and Design
Technical Committees 78-MCA and 51-ALC